Marek Berezowski

Is the brain capable to understand the brain?

AF138587

Marek Berezowski

Is the brain capable to understand the brain?

About algorithms, the brain, alien civilization, the theory of relativity and complex numbers

LAP LAMBERT Academic Publishing

Impressum / Imprint

Bibliografische Information der Deutschen Nationalbibliothek: Die Deutsche Nationalbibliothek verzeichnet diese Publikation in der Deutschen Nationalbibliografie; detaillierte bibliografische Daten sind im Internet über http://dnb.d-nb.de abrufbar.
Alle in diesem Buch genannten Marken und Produktnamen unterliegen warenzeichen-, marken- oder patentrechtlichem Schutz bzw. sind Warenzeichen oder eingetragene Warenzeichen der jeweiligen Inhaber. Die Wiedergabe von Marken, Produktnamen, Gebrauchsnamen, Handelsnamen, Warenbezeichnungen u.s.w. in diesem Werk berechtigt auch ohne besondere Kennzeichnung nicht zu der Annahme, dass solche Namen im Sinne der Warenzeichen- und Markenschutzgesetzgebung als frei zu betrachten wären und daher von jedermann benutzt werden dürften.

Bibliographic information published by the Deutsche Nationalbibliothek: The Deutsche Nationalbibliothek lists this publication in the Deutsche Nationalbibliografie; detailed bibliographic data are available in the Internet at http://dnb.d-nb.de.
Any brand names and product names mentioned in this book are subject to trademark, brand or patent protection and are trademarks or registered trademarks of their respective holders. The use of brand names, product names, common names, trade names, product descriptions etc. even without a particular marking in this work is in no way to be construed to mean that such names may be regarded as unrestricted in respect of trademark and brand protection legislation and could thus be used by anyone.

Coverbild / Cover image: www.ingimage.com

Verlag / Publisher:
LAP LAMBERT Academic Publishing
ist ein Imprint der / is a trademark of
OmniScriptum GmbH & Co. KG
Heinrich-Böcking-Str. 6-8, 66121 Saarbrücken, Deutschland / Germany
Email: info@lap-publishing.com

Herstellung: siehe letzte Seite /
Printed at: see last page
ISBN: 978-3-659-63760-5

Is the brain capable to understand the brain?
About algorithms, the brain, alien civilization, the theory of relativity and complex numbers

Marek Berezowski

For my Wife

Table of Contents

Chapter 1. Introduction

Although we shall not examine what the brain is and what processes take place in our brain, we shall still attempt to answer the question how it works, or, conversely, how it does not work. Is it only a very complex biological machine, or, maybe something more than that, and whether shall we ever be capable to understand it ? We shall prove that the brain understands without problems what machines cannot comprehend and will never be able to. Without problems we can read a text that does not exist. We shall tackle the issues of the theory of chaos and algorithms, but, first and foremost, we shall investigate the most complicated geometrical problem: the fractal. We shall try to derive mathematical equations to calculate Mandelbrot's fractal and examine how it behaves under dynamic conditions. We shall show how bicycle wheels draw fractals and check whether mathematics may create works of art and what this has in common with a chemical reactor. We shall ponder the question whether mathematics is omnipotent or is it rather a part of something more powerful. We shall consider possible forms of life in the Universe and think about ways of communication between people and aliens. Finally, we shall discuss the theory of relativity understandable to everyone and about the power of number *1*.

Chapter 2. A few words about fractals and chaos

Let us imagine a granary, into which two types of grains were poured: white and black. Let us assume that the boundary between the two types of grains is a certain line that divides the entire cubic capacity of the granary into two domains: one, where the white grains are, and the other one – containing the black ones, as shown in Fig.2.1

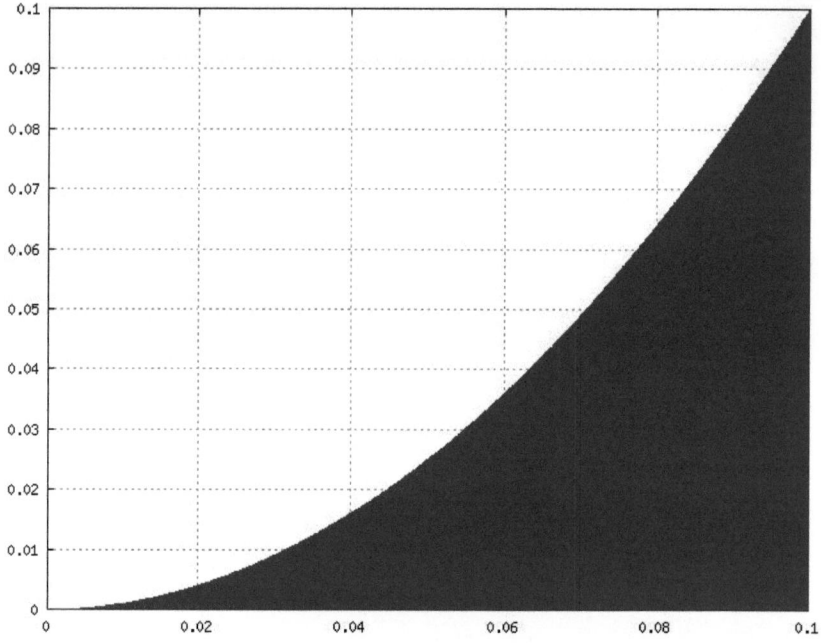

Fig. 2.1. The grains in the granary

Furthermore, let us assume that somebody has recklessly opened the door of the granary and strong wind has mixed the two types of grains. Undoubtedly, the new boundary is now not as clear as it was before. It is even not certain if the new boundary may be designated, as the wind has chaotically scattered the entire bunches of the grains all over the granary, leaving, at some spots, only bare ground, as shown in Fig. 2.2.

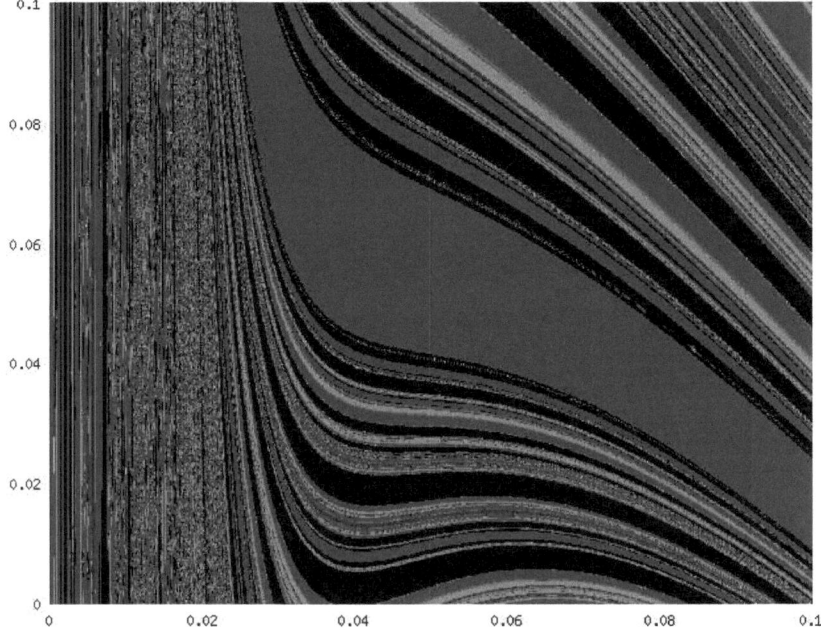

Fig. 2.2. The grains mixed up by the wind

Taking a closer look at the picture, we may notice that its smaller fragments are arranged in a similar shape, just as the whole surface. If, in turn, we analyze the small fragments, we will depict even smaller pieces, the shape of which is similar to the whole. Surely, the pieces are not of identical shape, but are only similar. One may ask the question: "What do they represent in terms of geometry?" In the beginning, before the impact of the wind, the granary was divided into two, clear-cut domains. But now the picture is different. The picture, in which smaller fragments are similar to bigger ones shall be referred to as a fractal. Let us remember that in the discussed example, it is the wind which is the author of the painting.

It should be noted that it was Richard Taylor, who in his article entitled " *Order in Pollock's Chaos* " [1] explained how the wind can "paint" fractal pictures. Taylor described a certain experiment. He spread out canvas on the branches of the trees and next to them, hooked a container with colorful paints, waiting for a strong wind. After the storm, he claimed that what the wind had "painted", by splashing the paints on the canvas, was a fractal: smaller fragments of the painting resembled the entire picture.

It is considered in science that the first person that "discovered" fractals in geometry is a mathematician, Benoit Mandelbrot, of Polish origin. He generated by means of a computer, a certain popular geometrical product, called Mandelbrot's fractal, named after him, to commemorate this breakthrough (Fig. 4.1).

Although his creation is generally considered as the most complicated geometrical figure that has ever been generated by man, it was derived from a relatively simple mathematical formula: $z \leftarrow z^2 + c$, where z and c are the so called complex numbers.

Fractals are present practically everywhere. So far, however, their strict definition has not been provided; nevertheless, they are characterized by the fact that- as a rule- they do not have an integral dimension. In classic Euclidian geometry the dimensions are always integral: a point has the dimension of 0, the line- the dimension of 1, the plane- 2, the solid – 3. In fractal geometry this may not be the case.

Chapter 3. Analytical deliberations on Mandelbrot's fractal

Let us reflect if the figure presented in Fig. 4.1, called Mandelbrot's fractal may be analytically designated, despite the fact that it is one of the most complex geometrical structures. In other words, is it possible to derive mathematical formulas that could enable strict, unambiguous and direct determination of all fragments of this figure? Let us attempt to answer this question. Because Mandelbrot's equation contains complex numbers, they are worth-mentioning at first. They are defined as:

$$z = z_r + iz_i \qquad (3.1)$$

where z_r and z_i are real numbers; whereas i is a fairly strange number, because when squared, it renders the value equal to -1:

$$(i)^2 = -1. \qquad\qquad (3.2)$$

Hence, it was labelled as an imaginary number.

Complex numbers are used in various calculations, for example in the case of oscillations. They may be graphically represented in the so called: "Argand plane" – see Fig. 3.1.

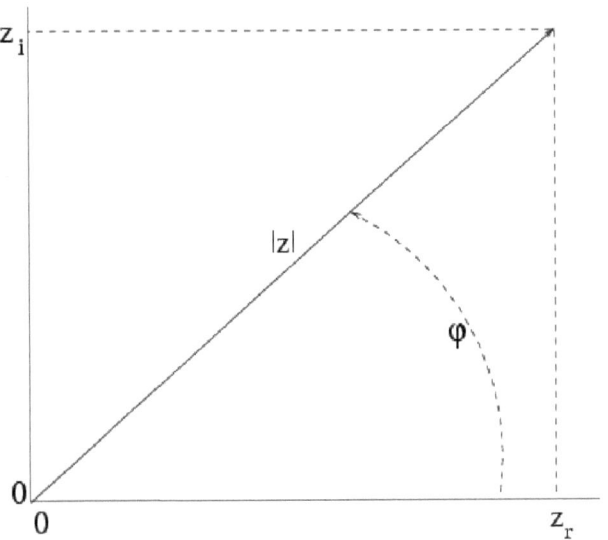

Fig. 3.1. Graphic representation of a complex number

The horizontal axis corresponds to the real part, whereas the vertical axis to an imaginary part of complex number z. By connecting point o with coordinates (z_r, z_i) with the origin of the coordinate system $(0,0)$ and applying Pythagoras theorem, the length of section $|z|$ may be easily designated from the following equation:

$$|z| = \sqrt{z_r^2 + z_i^2} \, . \qquad (3.3)$$

This section is inclined to axis z_r under a certain angle φ, which means that:

$$z_r = |z|\cos\varphi \, ; \quad z_i = |z|\sin\varphi \qquad (3.4)$$

Let us return now to Mandelbrot's fractal. Well, its full picture- as already mentioned before – is derived from a relatively simple formula:

$$z \leftarrow z^2 + c \qquad (3.5)$$

where variable z and constant c are complex numbers. A more detailed method will be demonstated below. With the use of equations (3.1) and (3.2), relation (3.5) may thus be written in a complex form, as follows:

$$z_r + iz_i \leftarrow (z_r + iz_i)^2 + c_r + ic_i \qquad (3.6)$$

or, in an equivalent form of two real equations:

$$z_r \leftarrow z_r^2 - z_i^2 + c_r \qquad (3.7)$$
$$z_i \leftarrow 2z_r z_i + c_i \, . \qquad (3.8)$$

Equations (3.5)-(3.8) are the so called : "recurrence equations", which means that the value of the left side substituted to the right side renders the successive value of the left side, etc. The numerical construction of Mandelbrot's set as shown in Fig. 4.1 is the following: For $z_r = 0$ and $z_i = 0$ arbitrary values of c_r and c_i are assumed, which, in accordance with relations (3.7)-(3.8) renders new values of z_r, and z_i. In turn, they provide the successive values, etc. If, after a big number of such recurrence steps, the length of section $|z|$, designated from equation (3.3), does not increase without limits, then on Argand plane, the point with coordinates (c_r, c_i) is set. If, however, the value of $|z|$ increases without limits (for example, it exceeds a certain very high number), no point is set on Argand plane. It has been proved that if $|z| > 2$, then, in the course of successive steps, the length of section $|z|$ increases. By repeating this procedure for the successive values of c_r and c_i a picture in the form of a cloud of points is obtained, as shown in Fig 4.1.

Let us now return to the question already posed in this paper, i.e. Is it possible to obtain Mandelbrot's set in an analytical way? In other words, is there a possibility of deriving mathematical equations, from which all fragments of the fractal can be calculated?

The answer requires some fundamental knowledge of the field of bifurcation of recurrence equations. Bifurcation is understood as a change in the type of a solution, for example, the change of the solution of a steady type for the solution of an oscillation type. Let us assume that for given values of c, i.e. for given values of c_r and c_i, relations (3.7)-(3.8) generate a constant sequence, that is, the sequence of invariable numbers: $|z|_k = (A,A,A...)$. In contrast, let us suppose that for a different value of c the above mentioned relations generate the oscillation sequence, or the sequence tends to infinity. Thus, there must exist a boundary value of c, which separates one type of the solution from another one. Analogically, as in the case of the granary. In mathematics, such boundary value is referred to as the bifurcation value, and the change in the solution type is called bifurcation. Let us now analytically designate the bifurcation value of constant c for Mandelbrot' set and, consecutively, mark its position on Argand plane. Yet, before this procedure, let us start with a simple example of linear bifurcation:

$$x \leftarrow ax \qquad (3.9)$$

which generates the sequence of numbers:

$$x_1 = ax_0$$
$$x_2 = ax_1 = a^2x_0$$
$$x_3 = ax_2 = a^3x_0$$

(3.10)

$$\cdot$$
$$\cdot$$

$$x_N = ax_{N-1} = a^N x_0$$

while x_0 is any arbitrary initial value of number x.

It is clear that if $|a|<1$, the successive values of variable x tend to zero. If, however, $|a|>1$, the successive values of variable x tend to plus or minus infinity, depending on the sign of parameter a. Value $|a|=1$ is, therefore, the bifurcation value, regardless of the fact whether a is a real or a complex number.

In the case where a is a real number, the above conclusion is obvious: raising the real number to a higher and higher power produces absolutely lower and lower result, if a is a number within the range : $-1<a<1$, but gives absolutely higher and higher result if a does not belong to this range.

If a is a complex number, it is possible – as we already know- to represent it in a graph in the form of the section with length $|a|$, inclined to the real axis under a certain angle. Thus, raising a complex number to a higher and higher power means decreasing or increasing the length of the section, depending on the fact whether its length was initially shorter or longer from the number *1*. Together with the raise of the power, the above mentioned angle is also increased, but this results only in the change of the inclination of the angle to the real axis, and does not affect the convergence of the sequence.

Relation (3.9) and its equivalent system of equations (3.10) are linear dependencies, as the left and right sides of the equations contain the terms describing straight lines. Mandelbrot's model (3.5) is non-linear, due to the fact that variable "z" is squared. Nevertheless, this model can also generate bifurcations. To designate them analytically, as before, the conditions should be set, for which the successive terms of sequence (3.5) cease to tend to a certain constant value (let us refer to it as z_s). In this manner, we will find the bifurcation point, and a set of such points will form the bifurcation line.

Let us first define the already mentioned value z_s. To achieve this, we may make use of relations (3.5) and (3.4), obtaining:

$$c = z_s - z_s^2 = |z_s|(\cos\varphi + i\sin\varphi) - |z_s|^2 (\cos\varphi + i\sin\varphi)^2. \quad (3.11)$$

The discussed method of designating the bifurcation point concerned a linear model. To apply the method to a non-linear model, it should be approximated by means of a straight line at point z_s (i.e. to perform the linearization at point z_s) and, continue the same bifurcation deliberations.
But, what does it mean to approximate a non-linear model by means of a straight line ? Let us consider the generalized form of the straight line equation:

$$y = \alpha x + \beta. \qquad (3.12)$$

Coefficient α next to variable x is a tangent of the inclination angle of this line to axis x. The straight line which approximates a curve at a certain point, is tangential at this point. This means that at the examined point the inclination angle of the curve to axis x is the same as the inclination angle of the tangential. Therefore, if we wish to write an equation of the straight line approximating the curve at a given point, at first, we must designate the tangent of the inclination line of the curve at this point. How to determine the value of this tangent? For the curve described by the generalized non-linear equation such as:

$$y = f(x) \qquad (3.13)$$

the searched tangent at point x_0 may be approximately determined as:

$$tg\varphi \approx \frac{f(x) - f(x_0)}{x - x_0} \qquad (3.14)$$

Clearly, the more value x is similar to the value of x_0, the more accurate the approximation (3.14). If x is infinitely closer to x_0, the designated tangent will be a derivative of function $f(x)$ at point x_0. Accordingly, to find out coefficient α in the straight line equation, approximating the curve at the set point, the value of the derivative of function $f(x)$ should be determined at this point. The straight line equation also contains coefficient β. Nonetheless, it is redundant, as only coefficient α is decisive for the bifurcation.

Thus, now we have at our disposal all the tools required for designating the bifurcation value of constant c in Mandelbrot's equation. The searched tangent of the approximating straight line is the derivative of the right side of equation (3.5) designated at point z_s. Hence:

$$\alpha = 2z_s. \qquad (3.15)$$

The analysis of linear equations (3.9)-(3.10) proves that the bifurcation holds for $|\alpha|=1$. Considering (3.15) this means that in the discussed case the bifurcation occurs for

$$|z_s| = \frac{1}{2}. \qquad (3.16)$$

Taking this value into account in (3.11), the complex equation is derived:

$$c = \frac{1}{2}(\cos\varphi + i\sin\varphi) - \frac{1}{4}(\cos\varphi + i\sin\varphi)^2 \quad (3.17)$$

which is equivalent to the system of real equations represented in the following parametric form:

$$
\begin{aligned}
c_r &= \frac{1}{2}\cos\varphi - \frac{1}{4}\cos 2\varphi \\
c_i &= \frac{1}{2}\sin\varphi - \frac{1}{4}\sin 2\varphi
\end{aligned}
\qquad (3.18)
$$

By changing the value of angle φ from 0 to 2π a complete bifurcation curve is obtained. It turns out that this curve is the boundary of the biggest part of Mandelbrot's fractal. In professional publications, it is often referred to as cardioid a due to its shape (similar to the shape of human heart), as shown in Fig. 4.1. Inside the domain designated by the *cardioid* the terms of sequence $|z|_k$ tend to the constant value. Outside the domain the terms of the sequence behave in a rapid manner, or, "run away" to infinity.

The boundaries of some other sub-domains of Mandelbrot's fractal may be designated in a similar manner.

Let us assume that this time the bifurcation applies to the oscillation solution: $|z|_k = (A,B,A,B...)$. In such case, constant point z_s should not be designated from a single recurrence $z \to F(z)$, but from the so called double circulation $z \to F[F(z)] = G(z)$, entailing two constant points A and B. The boundary of the stability of the circulation may be designated analogically, as it was discussed above, i.e. by the analysis of the linear approximation. In such approximation, coefficient α is the derivative of function G determined at any fixed point. If the recurrence system loses its stability, both point are destabilized at the same time.

For Mandelbrot's model, function G assumes the following form:

$$G = (z^2 + c)^2 + c. \qquad (3.19)$$

and its derivative:

$$\alpha = \frac{dG}{dz} = 4z(z^2 + c). \qquad (3.20)$$

So, the boundary condition of the convergence of sequence $|z|_k = (A,B,A,B...)$ is:

$$|\alpha| = 4|z_s(z_s^2 + c)| = 1. \qquad (3.21)$$

The fixed points may be designated from the double circulation equation as:

$$z_s = \left(z_s^2 + c\right)^2 + c \ .$$

(3.22)

Combining (3.21) and (3.22) the equations determining the bifurcation of the rapid oscillation sequence may be obtained. Likewise, they have a parametric character and the following form:

$$c_r = \frac{1}{4}\cos\varphi - 1$$
$$c_i = \frac{1}{4}\sin\varphi$$

(3.23)

By changing the value of angle φ from 0 to 2π, the boundary of the smaller domain from Fig 4.1 is obtained. The discussed oscillation sequence is convergent in the domain limited by the boundary.

Yet, if someone believed that the analytical designation of the entire Mandelbrot's set, i.e. all its boundaries was achievable, they would be wrong. This is impossible, first and foremost, due to the fact that the number of such boundaries is infinite. Our ability to perform mathematical transformations does not even enable the designation of the forth boundary, as this would imply the necessity of determining the roots of higher order polynomials, which is analytically impossible. Readers more interested in this field should refer to other publications [2,3,4].

The deliberations contained in this chapter give the grounds for further discussion on Mandelbrot's dynamic equation with delay.

Chapter 4. Analysis of the dynamics of Mandelbrot's set with delay

There are many works devoted to the Mandelbrot's equation and set. As we know from chapter 3, in order to recall it, Mandelbrot's equation constitutes a recurrent form

$$z_{k+1} = z_k^2 + c \qquad (4.1)$$

in which the variable z and constant c are complex numbers. In the result of recurrence (4.1) one obtains, for $z_0 = 0$, the sequence z_k, which is convergent or divergent, depending on the value of constant c. Confirming the convergence of this sequence by means of the point on the complex plane $c(c_r, c_i)$, results in the fractal structure, named – after its discoverer – Mandelbrot's fractal (Fig. 4.1).

In spite of the fact that this structure is extremely complicated, it is possible to determine analytically its principal fragments. Viz., it may easily be shown that the edge of its largest area is a set of Hopf bifurcation points satisfying the condition $|z| = \frac{1}{2}$ (see chapter 3). This set may be described by the complex parametric equation of the form

$$c_H = \frac{1}{2} e^{i\varphi} - \frac{1}{4} e^{2i\varphi} \qquad (4.2)$$

where parameter φ is an argument of the complex variable z (full line in Fig.4.1). Also quite easily one can determine the edge of solutions of two-period oscillations

$$z_{k+2} = (z_k^2 + c)^2 + c \qquad (4.3)$$

which satisfies the condition

$$4 \,|\, z(z^2 + c) \,| = 1. \qquad (4.4)$$

The complex parametric equation has the form

$$c_{2H} = \frac{1}{4}e^{i\varphi} - 1 \qquad (4.5)$$

where parameter φ is the argument of complex variable (Fig. 4.1). The above equations were discussed in chapter 3 in the form of real equations (3.18) i (3.23).

In a similar way one may also determine the edge of three – period solutions, for which the corresponding complex variable fulfils the condition

$$8 \,|\, z(z^2 + c)[(z^2 + c)^2 + c] \,| = 1 \,. \qquad (4.6)$$

The relevant discussion was carried out a.o. in [2,3].

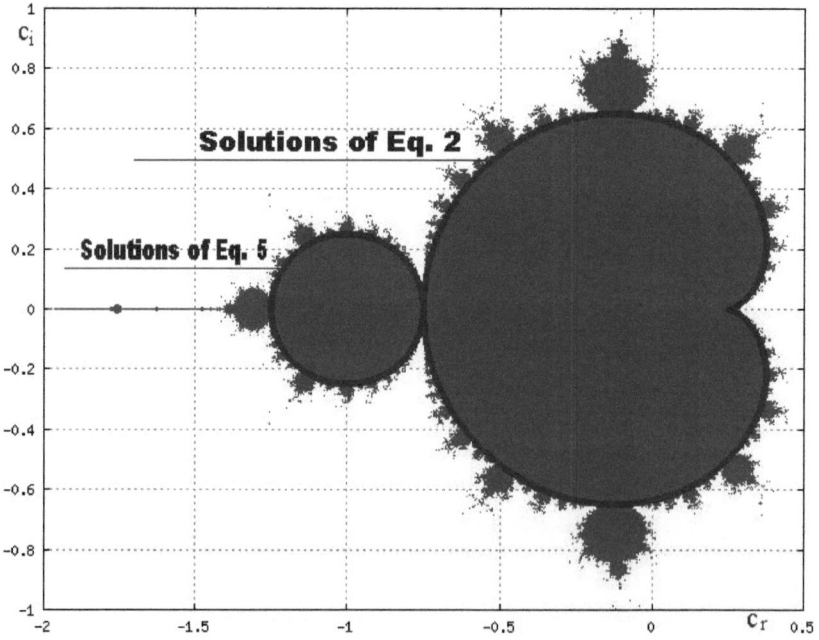

Fig. 4.1. Bifurcation curves (thick lines)

Let us assume that the general recurrent form

$$z_{k+1} = F(z_k) \qquad (4.7)$$

constitutes a mathematical model of some dynamic system with delay. Hence, such a system may be described by an equivalent continuous equation of the form

$$z(t) = F[z(t-t_0)] \qquad (4.8)$$

where t_0 is a certain delay time. If, additionally, given dynamic system is characterized by a significant inertion, the corresponding derivative with respect to time must appear in Eq.(4.8) [4,5], which leads to the relationship of the form

$$\sigma \frac{dz(t)}{dt} + z(t) = F[z(t-t_0)] \qquad (4.9)$$

Let us assume in our considerations that the mathematical relationship of Mandelbrot is connected with a certain continuous dynamic system comprising delay and inertion, the latter characterized by the dynamic capacity σ. The Mandelbrot's equation (4.1) extended in such a way has the form:

$$\sigma \frac{dz(t)}{dt} + z(t) = z^2(t-t_0) + c. \qquad (4.10)$$

Introducing the normalized time $\tau = \frac{t}{\sigma}$ one obtains the final equation

$$\frac{dz(\tau)}{d\tau} + z(\tau) = z^2(\tau - \tau_0) + c \qquad (4.11)$$

In order to analyze the behavior of solutions of this equation, use has been made of the above – mentioned Mandelbrot's algorithm, which generates the fractal set. To this aim there was constructed the map of stable and unstable solution of Eq.(4.11) as a function of successive values of complex constant c

under the assumption $z(0) = 0$. For $\tau_0 = 0$ Eq.(4.11) reduces to the differential form without delay. In consequence, its solution cannot be of chaotic character and therefore the corresponding region cannot display the fractal character. Another extreme case is $\tau_0 = \infty$, which means that the form (4.11) is equivalent with the recurrent form (4.1) of Mandelbrot. In the chapter presented some exemplary computations were carried out under the assumption $\tau_0 = 10$; the corresponding image is shown in Fig. 4.2.

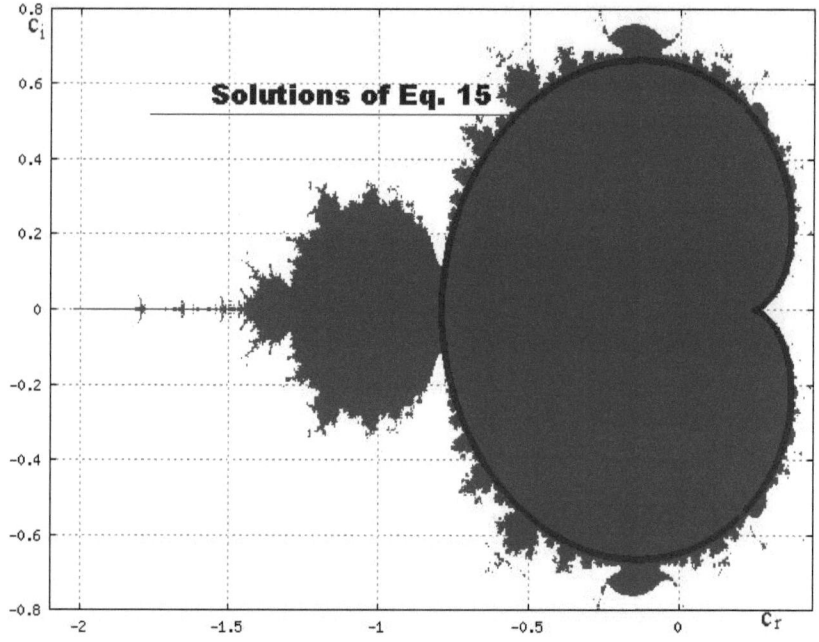

Fig. 4.2. Image for small time simulation

It may be seen that this image is very similar to the original Mandelbrot's fractal form (Fig. 4.1), which seems to be a quite natural result. This topic will be discussed later. Similarly as in the case of the model (4.1), taking into consideration Eq.(4.11), one can obtain here the analytical relationship, which determines the edge of stable stationary solutions of the model (full line in Fig.4.2). Namely, the linear approximation of Eq.(4.11), determined in the vicinity of stationary solution, is

$$\frac{du}{d\tau} + u = 2z_s u(\tau - \tau_0) \qquad (4.12)$$

where $u = z - z_s$, z_s being the stationary solution of relation (4.11). The characteristic equation of relationship (4.12), written in the frequency form, is

$$1 - \frac{2|z_s|e^{i\varphi}}{\sqrt{1+\omega^2}} e^{-iarctg\omega} e^{-i\omega\tau_0} = 0. \qquad (4.13)$$

Using Nyquist criterion one obtains the boundary of stability of Eq.(4.11) in the form

$$|z_s| = \frac{\sqrt{1+\omega^2}}{2}. \qquad (4.14)$$

Making use of the definition of the stationary state ($\frac{dz}{d\tau} = 0$) one obtains finally the parametric relationship determining the edge of stationary solutions of Eq.(4.11) in the form

$$c_H = \frac{\sqrt{1+\omega^2}}{2} e^{i\varphi} - \frac{1+\omega^2}{4} e^{2i\varphi} \qquad (4.15)$$

where parameter φ is an argument of complex variable z_s; the circular frequency ω is determined from the relation

$$\omega\tau_0 + arctg\omega = \varphi. \qquad (4.16)$$

The regions outside the main part of Mandelbrot's set refer to the multiperiodic oscillations [4,6]. It comes out that during the formation of the structure as in Fig.4.2 the magnitude of these external regions diminishes with the increase of the time of simulation of eq.(4.11) (Fig.4.3).

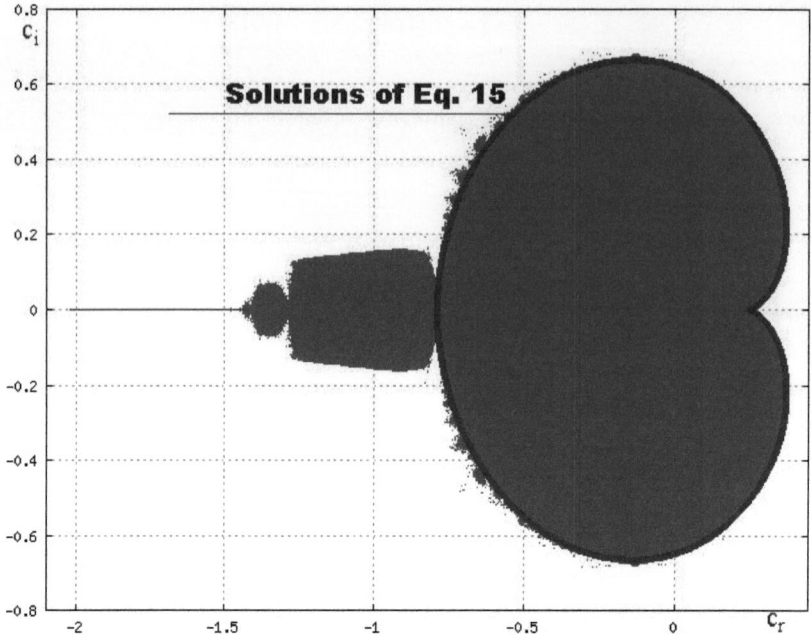

Fig. 4.3. Image for large time simulation

This means that for the infinitely long simulation time all regions vanish, except the main one. This results from the fact that the oscillation solutions of Eq.(4.11) (except the case $c_i = 0$) are unstable. So, the geometric structure generated according to the above – mentioned algorithm consists exclusively of the main region, which characterizes the stable stationary solutions of this equation and of a single line, which starts to the left of this region from the point of the coordinates (-0.791;0) (Fig.4.4).

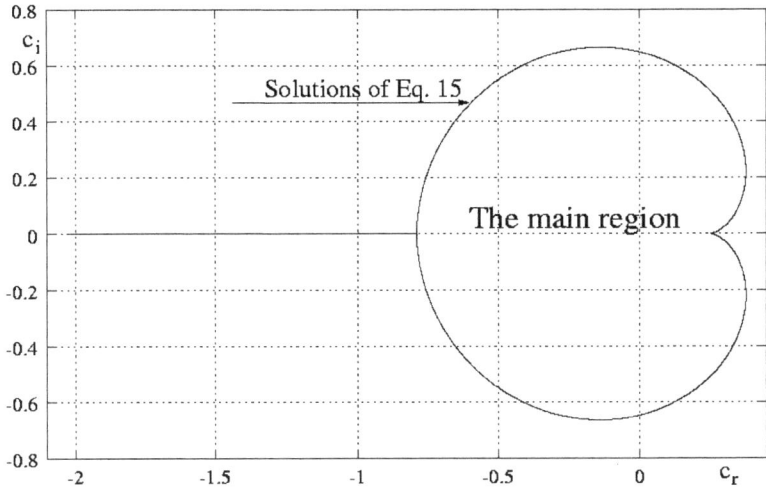

Fig.4.4. Image for infinite time simulation

This line, as the only exception, constitutes the set of stable periodic and chaotic solutions of Eq.(4.11), as shown in the Feigenbaum's diagram of steady states (Fig.4.5). Similar results are obtained for any arbitrary delay time $\tau_0 < \infty$.

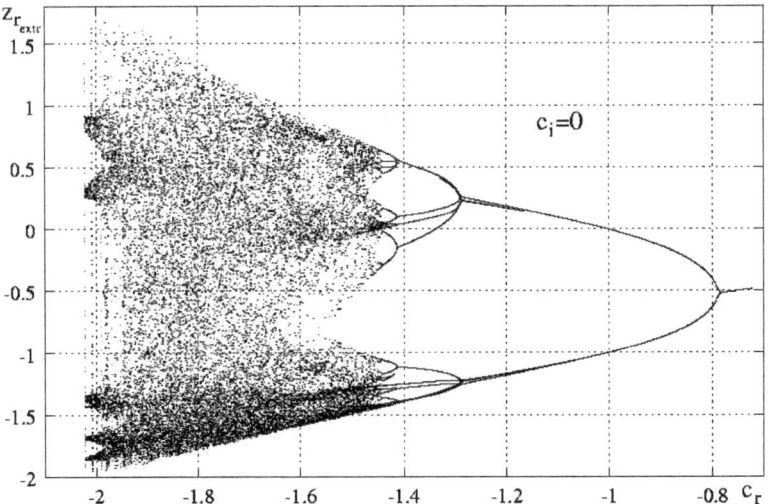

Fig.4.5. Feigenbaum diagram

So, the Mandelbrot's equation (4.1), as a particular case of the form (4.11) is an exception, which generates the fractal structure. Viz., if in this equation the inertial term appears, the fractal vanishes. This phenomenon may be investigated, analyzing the time series. First, however, let us discuss more precisely the edge of the region from Fig.4.5, i.e. the set of points of Hopf bifurcation. To each of these points a certain value of the angle φ corresponds, and thus - the circular frequency ω. The phase displacement, due to the right - hand side of Eq.(4.11) depends solely on φ and τ_0. It results from Eq.(4.16) that the circular frequency ω at the Hopf bifurcation points also depends exclusively on φ and τ_0. This means that for the fixed values of φ and τ_0 the frequency of oscillations does not depend on the value of the complex variable c. Since the real and imaginary parts of the complex variable c_H at Hopf bifurcation point are expressed as:

$$c_{Hr} = \frac{\sqrt{1+\omega^2}}{2}\cos(\varphi) - \frac{1+\omega^2}{4}\cos(2\varphi) \qquad (4.17)$$

$$c_{Hi} = \frac{\sqrt{1+\omega^2}}{2}\sin(\varphi) - \frac{1+\omega^2}{4}\sin(2\varphi) \qquad (4.18)$$

the tangent of the argument of this constant ($k = \frac{c_{iH}}{c_{rH}}$) depends only on φ. Hence, at each point $c(c_r;kc_r)$ the frequency of oscillations is the same, no matter the value of c_r. Let us assume, as an example, $\varphi = 2.517$, which - according to Eqs (4.16)-(4.18) – corresponds to $\omega = 0.2292$. Thus, the period of oscillations is equal to $T = \frac{2\pi}{\omega} = 27.417$. The numerical simulation of the time series $|z(\tau)|$ confirms the above analytical calculations, which is shown in Fig.4.6.

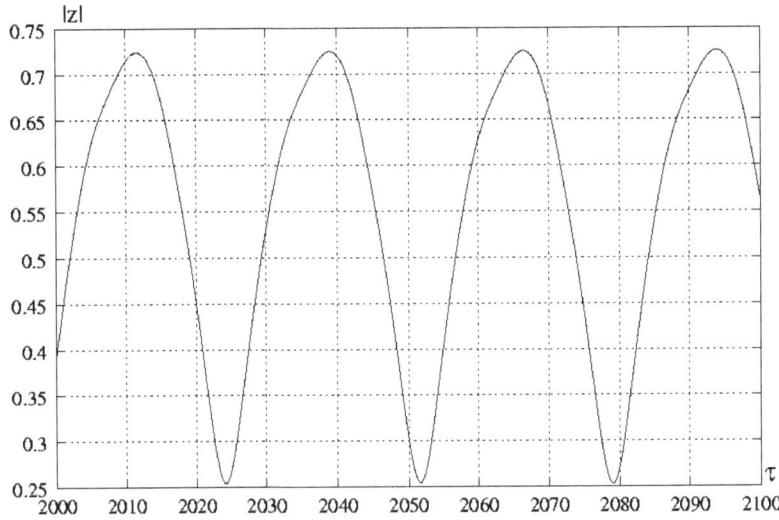

Fig.4.6. Time story for small time simulation

However, the result of the simulation is only approximately true. Namely, at the increase of the simulation time one finds that this time series is unstable (Fig.4.7).

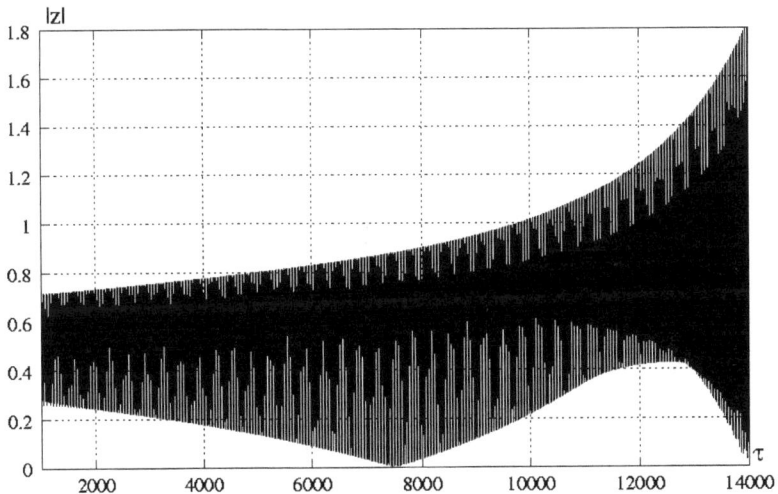

Fig.4.7. Time story for large time simulation

The consequence of it is the vanishing of the fractal structure of the image with the increase of simulation time. It should be added that the reverse phenomenon does not have to occur. This denotes that the removal of the time derivative from the complex differential equation with delay and thus the change of its form from the continuous form to the logistic one does not have to cause the generation of the fractal sets from this form. An example here is Eq.(4.8) analyzed in [5].

Chapter 5. Bike wheels paint fractals

The dynamics of a mechanical device consisting of n wheels of different radii and rotating at different speed was tested. On the rim of a given wheel rotating with specific speed another smaller wheel is mounted, rotating at a higher speed. (Fig. 5.1).

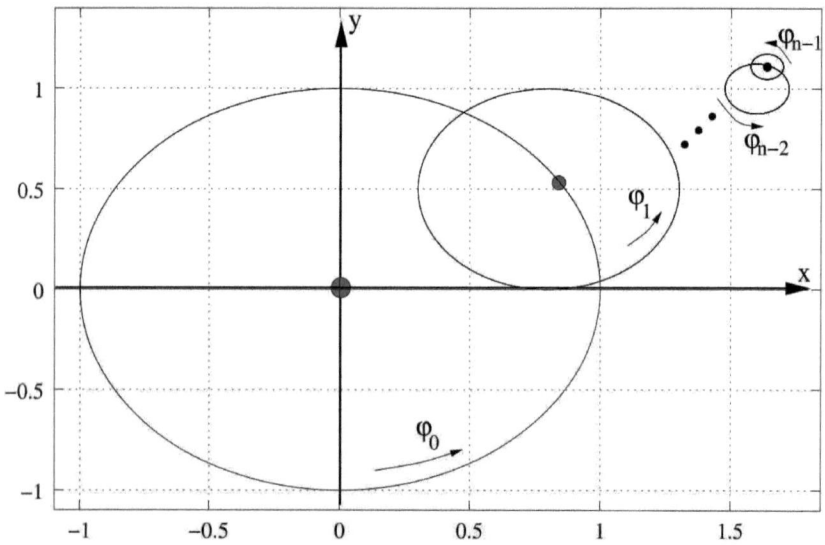

Fig. 5.1. Conceptual diagram of the system of wheels

Accordingly, the particular wheels rotate not only at their own speed but are also driven by the motion of all the wheels that are bigger. On the rim of the last (the smallest) wheel a scriber is mounted, leaving traces on the plane. It turns out that even with a relatively small number of the wheels, despite a non-chaotic motion of the scriber, the trace may be very complex and may create a fractal figure [7]. It is an interesting fact that the speed of the scriber is constantly changing, both in terms of its value and direction, although the angular speeds of the wheels are constant. In an extreme case, when the number of the wheels approaches infinity, the speed of the scriber changes at each moment. Such behavior is similar to Brownian motion.

As already mentioned in the Introduction, let us consider a mechanical device consisting of n different wheels. On the rim of the k-th wheel axis $(k+1)$-th is mounted (Fig.5.1). The wheels rotate at specific speeds in the counter-clockwise direction. Let us assume that the rations between the radii and the angles of rotation of the wheels are constant and equal to:

$$q = \frac{r_k}{r_{k+1}} = \frac{\varphi_{k+1}}{\varphi_k} > 1. \qquad (5.1)$$

The scriber mounted on the rim of the last wheel leaves traces on the surface. The position of the scriber is determined by its coordinates:

$$x' = \sum_{k=0}^{n-1} r_k \cos(\varphi_k) = r_0 \sum_{k=0}^{n-1} \frac{1}{q^k} \cos(q^k \varphi_0) \qquad (5.2)$$

$$y' = \sum_{k=0}^{n-1} r_k \sin(\varphi_k) = r_0 \sum_{k=0}^{n-1} \frac{1}{q^k} \sin(q^k \varphi_0) . \qquad (5.3)$$

In terms of a complex notation, the above equations may be expressed as:

$$R' = \sum_{k=0}^{n-1} r_k e^{i\varphi_k} = r_0 \sum_{k=0}^{n-1} \frac{1}{q^k} e^{iq^k \varphi_0} \qquad (5.4)$$

where R' is the complex radius-vector.

On the grounds of the above relationships it is easy to indicate that for $n=2$ and $q=2$ the scriber marks the line the length of which is $L=8r_0$. For $r_0=1/2$ the line demarcates the area of identical size and shape as the biggest Mandelbrot fractal, often referred to in literature as "*the heat curve*" [2,3,4].

By introducing the standardized coordinates and the reduced notation of the angular variable:

$$R = \frac{R'}{r_0}; \quad t = \varphi_0 \tag{5.5}$$

equation (4) is transformed into:

$$R = \sum_{k=0}^{n-1} \frac{1}{q^k} e^{iq^k t} . \tag{5.6}$$

This dependence is analogical to Weierstrasse function [8,9], the only difference being that the derivative of Weierstrasse function is infinite; whereas, in the discussed model, the derivatives are finite. As R changes in a continuous mode, the trajectory of the motion of the scriber is also continuous. However, the rate of the change of the radius is discontinuous for $n \to \infty$:

$$\frac{dR}{dt} = i \lim_{n \to \infty} \sum_{k=0}^{n-1} e^{iq^k t} . \tag{5.7}$$

For $n = \infty$ the discontinuity occurs for each value of angle t. To prove this, let us consider the following boundary:

$$\varepsilon = i \lim_{\Delta t \to 0} \sum_{k=0}^{\infty} \left(e^{i\left(q^k t + q^k \Delta t\right)} - e^{iq^k t} \right). \tag{5.8}$$

As index k is directly equal to infinity, and the increase of angle Δt only approaches zero, assuming that $q>1$ – quotient $q^k \Delta t$ is equal to infinity. This means that for each value of t boundary ε is different from zero, which proves that the derivative presented above is discontinuous for each value of t. In practice, when $n < \infty$ this phenomenon is important to $\Delta t > q^{1-n}$. Thus, for example, for $n=20$ and $q=2.5$, $\Delta t > 0.00000003$. This means that the loss of the discontinuity of the derivative is observable in the course of the measurements of the motion of the biggest wheel only at intervals shorter than 30 nano-degrees. Accordingly, in practice the system is subject of incessant rapid changes of the speed of the scriber motion, both in terms of its value and direction. It may be stated that the scriber is continuously driven, creating, in consequence, very complex fractal figures on the surface.

To set an example, two cases were considered: $q=2.5$ and $q=5$. For each of the cases $n=20$ was assumed. Accordingly, the curve created by the scriber mounted at the 20th wheel for $q=2.5$ is shown in Fig.5.2.

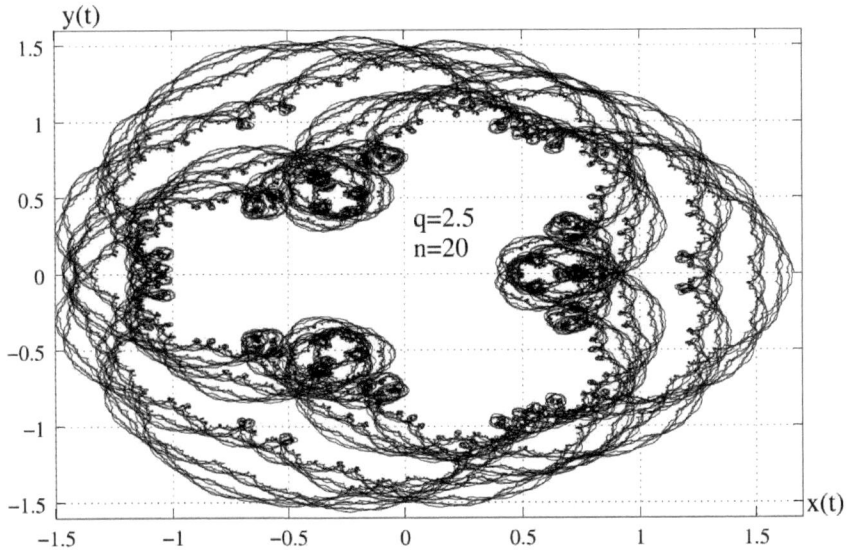

Fig. 5.2. Phase diagram of the scriber on the phase plane. $q=2.5$, $n=20$

The presented geometrical structure is very complex, but the images of its fragments point to a fractal form (Fig. 3 and 4) [7].

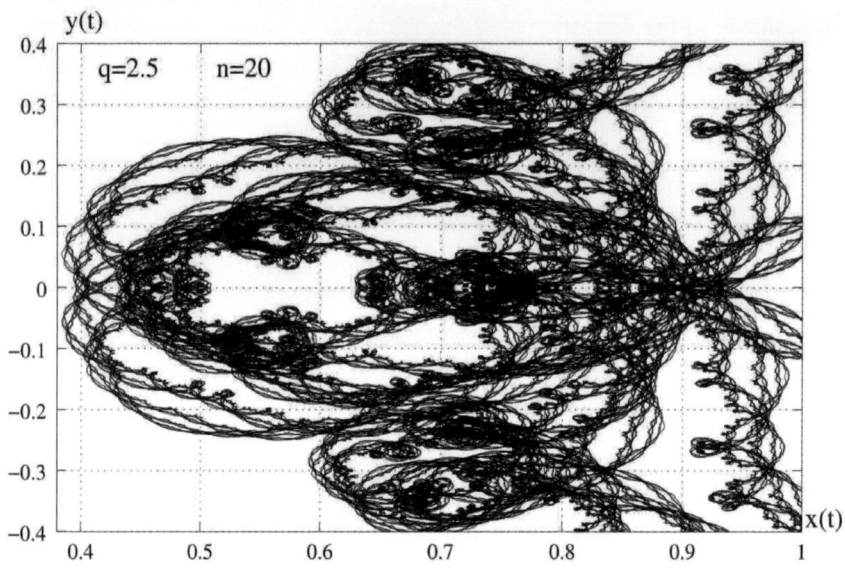

Fig. 5.3. Fragment of Fig. 5.2

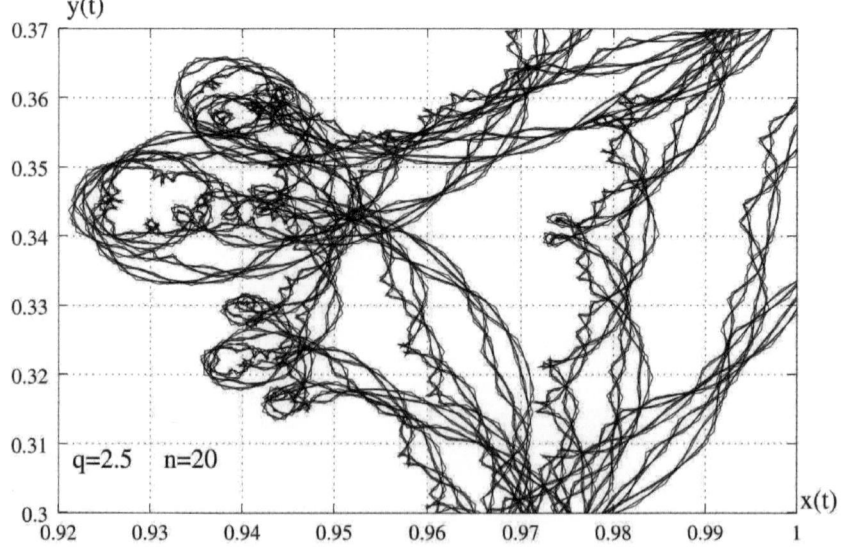

Fig. 5.4. Fragment of Fig. 5.3

Coordinates x and y are normalized in accordance with the dependence:

$x = \dfrac{x'}{r_0}$, $y = \dfrac{y'}{r_0}$. Periodic changes of the length of radius-vector $|R|$ are illustrated

in Fig.5.5, whereas in Fig.5.6 the speed of $\left|\dfrac{dR}{dt}\right|$ changes is shown.

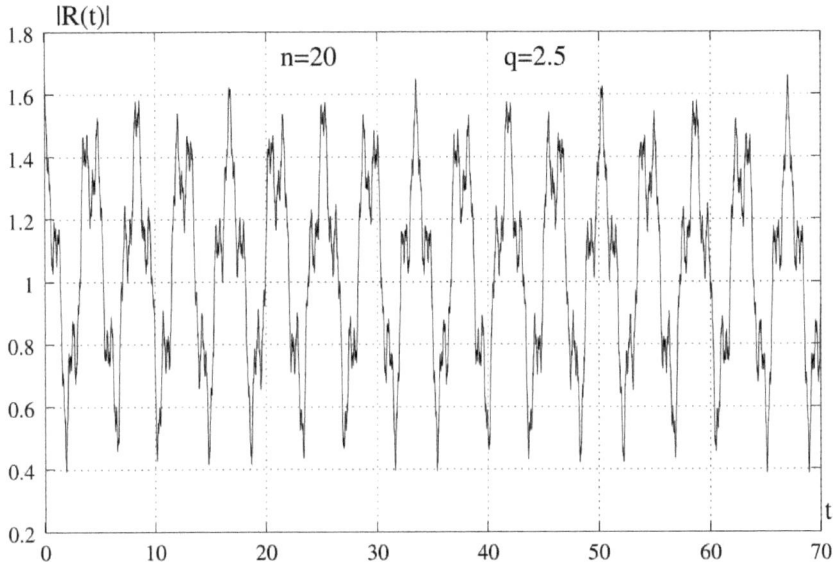

Fig. 5.5. The dependence between the radius vector and the angle of rotation of the biggest wheel. $q=2.5$, $n=20$

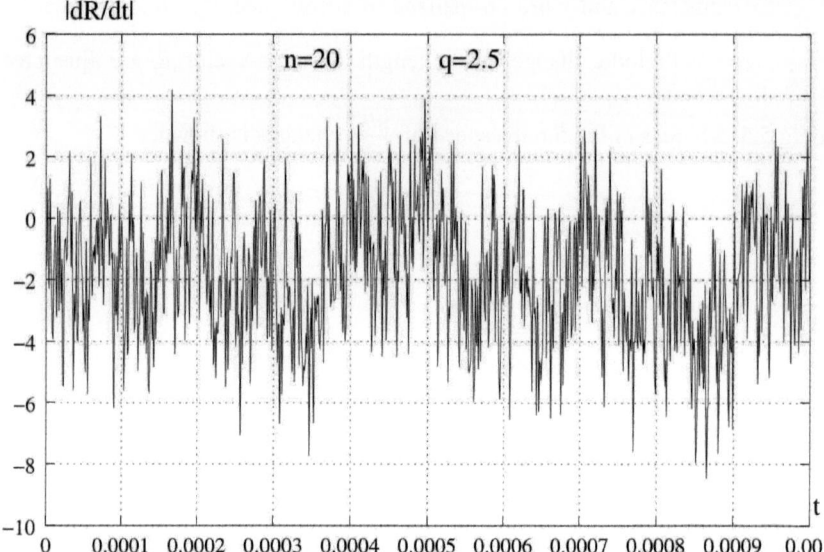

Fig. 5.6. Dependence between the rate of the changes of the radius vector and the angle of rotation of the biggest wheel. *q*=2.5, *n*=20

The graphs certify the continuity of *R(t)* and the discontinuity of its derivative in practice. In an extreme case, for each value of angle *t* a rapid change of speed occurs. The zigzags visible on the phase plane are a consequence of the above-mentioned discontinuity (Fig.5.7).

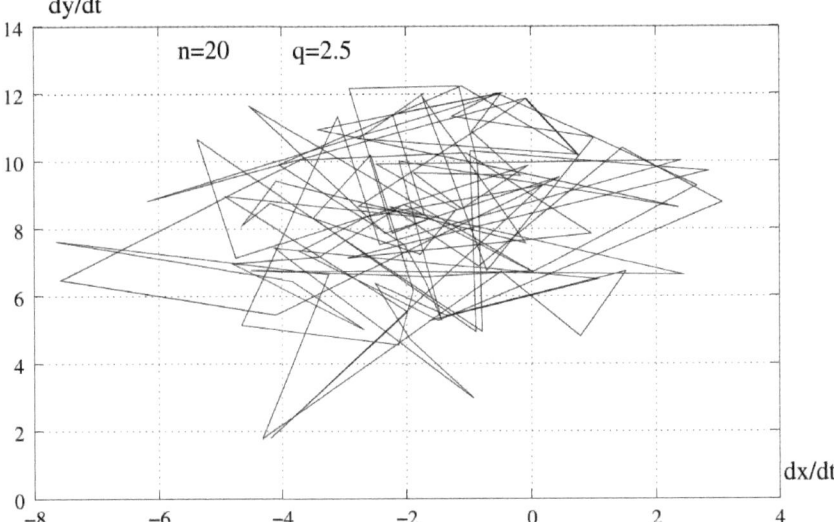

Fig. 5.7. Phase trajectory of the rate of changes in the coordinates of the scriber.
q=2.5, n=20

When $n = \infty$ the zigzags occur at each point of the plane. Such trajectory resembles Brownian motion, where, as commonly known, the motion of basic particles is also zigzag-like at each point of the surface [10,11].

It should be emphasised that even though the motion of the scriber is not of a chaotic nature, yet it is very sensitive to changes in ratio $\frac{\varphi_{k+1}}{\varphi_k}$. Thus, in Fig.5.8 the thick line represents the case when $q = 2.5$, whereas the thin line the case of $q_\varphi = \frac{\varphi_{k+1}}{\varphi_k} = 2.499$.

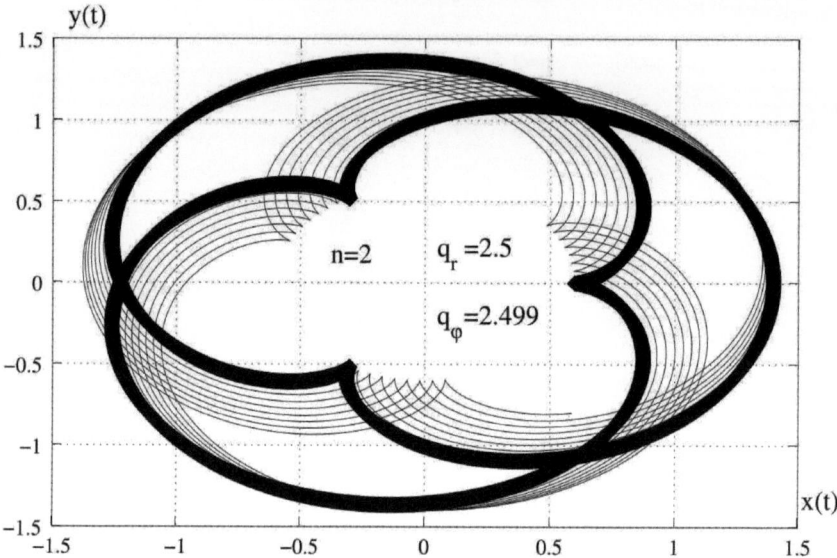

Fig. 5.8. The sensitivity of the system to changes in the angular speed ratios. The thick line: q=2.5. The thin line: $q_r = 2.5$, $q_\varphi = 2.499$

Both graphs refer to n=2. Yet, the change of the value of q_r, given the same q_φ value, does not affect significant changes in the course of the trajectory. In Fig.9 the phase trajectory of the scriber motion was shown for q=5.

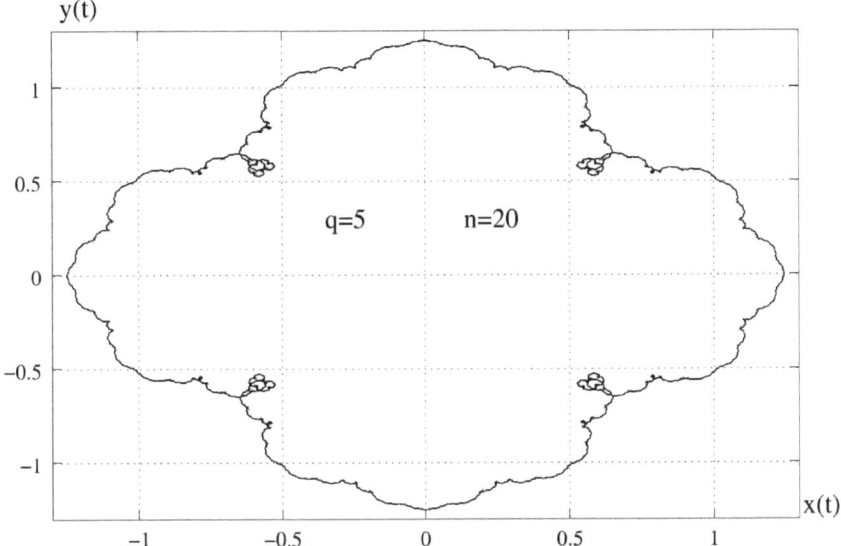

Fig. 5.9. Phase diagram of the motion of the scriber on the plane. $q=5$, $n=20$

It may be observed that this motion is not as complex as in the previously discussed case, nonetheless, the derived graph has a fractal nature, as substantiated by the fragments presented in Figures 10 and 11. After scaling, the Figures are identical.

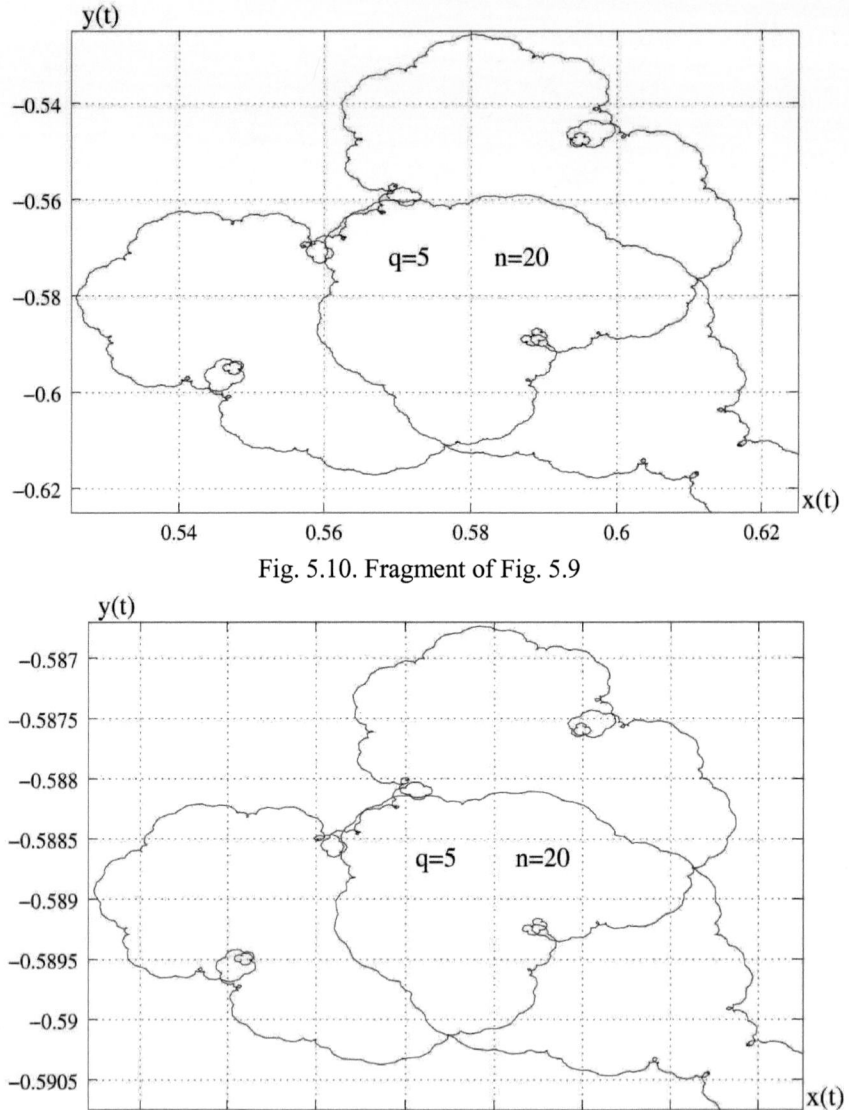

Fig. 5.10. Fragment of Fig. 5.9

Fig. 5.11. Fragment of Fig. 5.10

In Fig.12 changes in the information system entropy are indicated, depending on q.

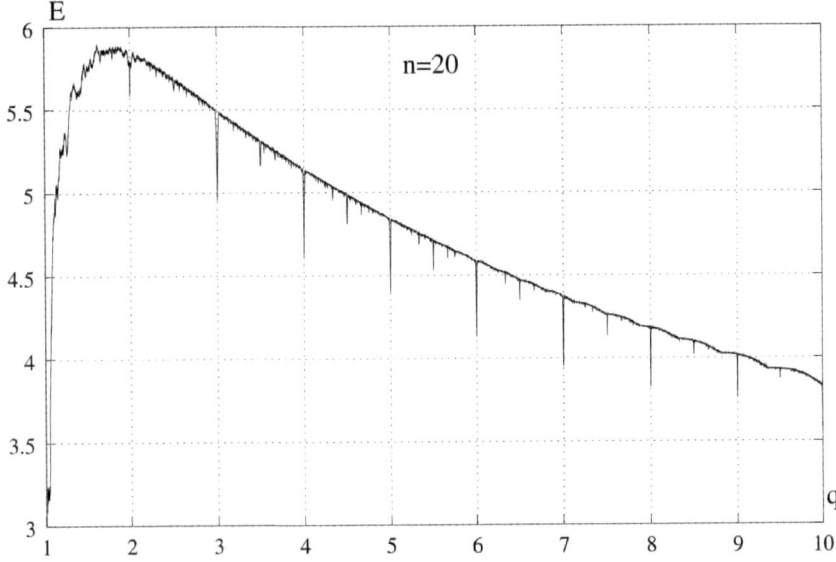

Fig. 5.12. Dependence between the information system entropy and q ratio

They also have a fractal nature. The values of the entropies were calculated from the following formula:

$$E = -\sum_{i=1}^{N} p_i \lg_2 p_i \qquad (9)$$

where p_i is the probability of the occurrence of a definite value of the length of radius $|R/$ [7]. As seen in the above graph, most information is rendered by the system for $q=1.8$.

Concluding of this chapter

A mathematical and numerical analysis conducted within the framework of the chapter was focused on the dynamics of a mechanical system consisting of *n* wheels rotating at constant speed around their axes. The smaller wheels were mounted on the rim of bigger wheels- see Figure 5.1. It was assumed that the ratios between the radii and the angles of rotation of the successive wheels are constant. There is a scriber mounted on the rim of the smallest wheel. In the outcome of the analysis, very complex fractal graphs were derived. The behavior of the speed of the scriber is very interesting, as its components, represented on the phase plane, resemble Brownian motion (Fig.5.7).

Although the changes in the length of the radius vector have a continuous character (Fig.5.5), yet the course of the changes is not even. With an infinite number of the wheels the roughness of the graph occurs at each of its points. A spatial form of this phenomenon is shown in Fig.5.13.

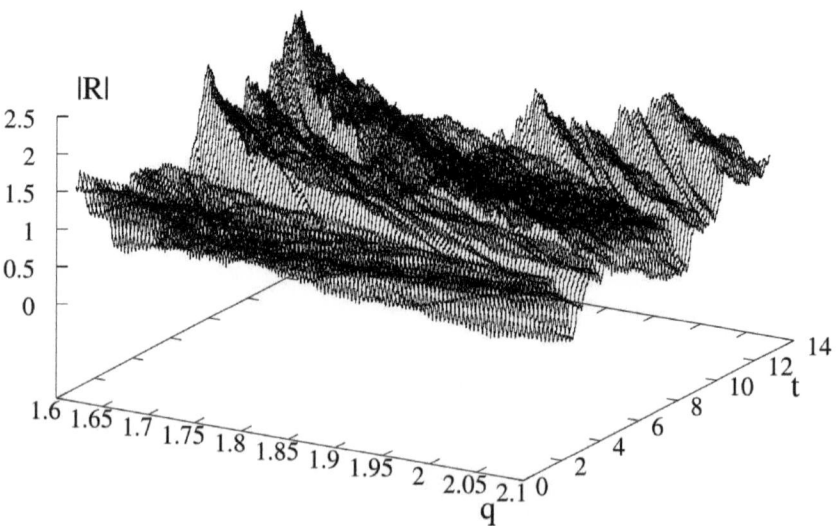

Fig. 5.13. Visualization space of the changes of the radius vector

It is difficult to refrain from the impression that natural biological structures have the same character. It is sufficient to have a look at afforested mountain slopes, rocky or non-stony. Therefore, a conclusion may be drawn that in the natural environment we can also witness the phenomenon of common roughness.

Chapter 6. What do art and a chemical reactor have in common?

As mentioned before, in 2000 Richard P. Taylor, Professor of Physics at The State Oregon University, published an intriguing article [1]. Its main focus was the mathematical and computer analysis of selected works of painter Jackson Pollock.

In his article Taylor indicated that the artistic structures created by Pollock had a fractal character. He proved that the fragments of a painting were similar to the whole and, in practice, had the same dimensions as the whole of the painting. The feature typical of fractals is the fact that a fragment of a certain structure, magnified by a proper number of times, resembles the whole and has, approximately, the same dimensions as the whole. Furthermore, Taylor proved that the dimensions of the structures created by Pollock increased over the years from $D=1,12$ to $1,90$. This means that they became more and more complex in his successive canvases.

The fractional value of the dimension denotes that we are dealing with space that is not entirely filled up, so with something in- between the first and the second dimension. So, an example of a fractal with the dimensions within the range from 0 to 1 is the so called Cantor set consisting of the sections of zero length ($D=0.631$; Fig.6.1).

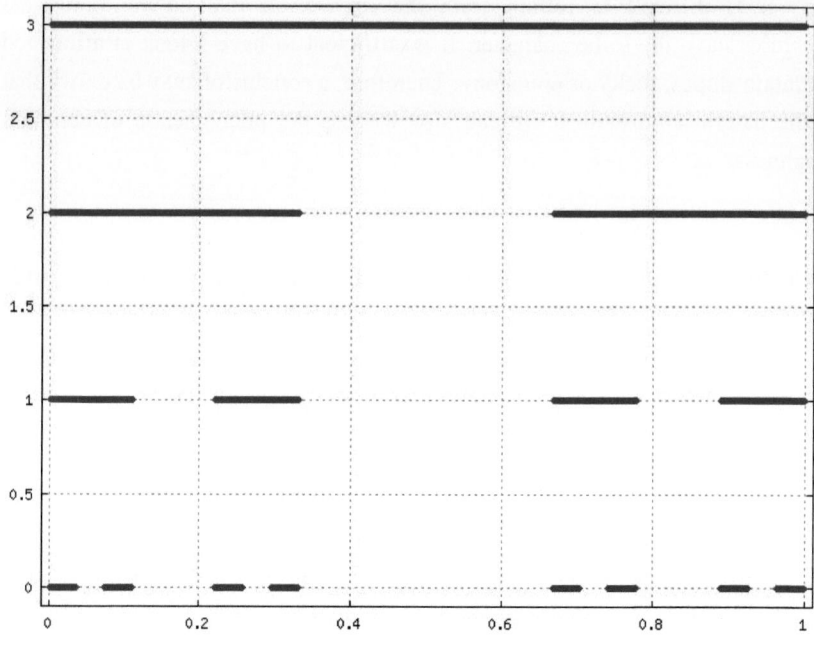

Fig. 6.1. Cantor set

For the case where the value of dimension D is within the range from 1 to 2, a good example is the of infinite number of triangles with zero field (for instance: Sierpinski triangle, $D=1.585$, Fig. 6.2) whereas in the case of D within the range from 2 to 3, a good example is the set of an infinite number of cubes with zero volume (for instance: Menger sponge, $D=2.727$, Fig.6.3).

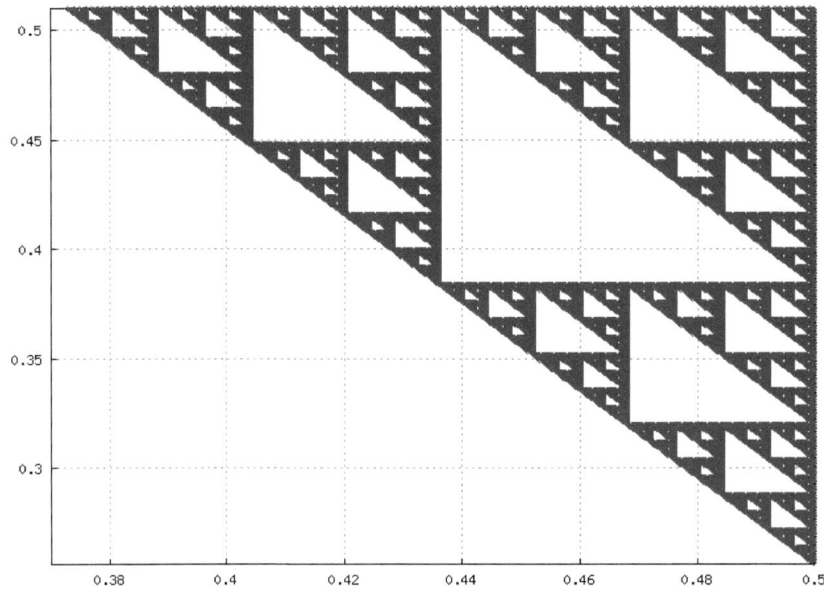

Fig. 6.2. Sierpinski triangle

Fig. 6.3. Menger sponge

While determining the dimensions of Pollock's works, Taylor applied the so called: "lattice method. Generally speaking, the method involves dividing a certain painting, or its fragment, into N least squares and next calculating the number of squares N_s, in which characteristic elements of the painting appear. Then, the dimension of the investigated structure may be approximately determined from equation:

$$D = 2\frac{\lg N_s}{\lg N}.\qquad\qquad (6.1)$$

We encounter fractal structures in almost every area of our daily life. They are practically present everywhere in the world. While looking at a photograph of a stone placed against neutral background, we cannot claim with certainty if what we can see is a stone or a huge mountain (this effect is sometimes used by stage designers). A branch of a tree, magnified by an appropriate number of times, may resemble the entire tree. Likewise, the clouds, snowflakes, shorelines, lightning, the nervous system, the circulatory system, broken pane, etc. Thus, it may resolutely be claimed that fractals are geometrical formations which surround us, as in nature there are no perfect segments, circles or cubes.

Does the modeling of a complex fractal structure have to be mathematically complicated? Certainly not. A good example is the already discussed Mandelbrot's fractal, which, despite of being a structure of infinite complexity, may be represented by means of a simple recurrence equation.

One of the problems in which the author of this paper is deeply involved is the analysis of the dynamics of chemical reactors, i.e. apparatuses in which chemical reactions occur. Strictly speaking, the analysis does not concern physical devices, but their mathematical models. It turns out that chemical reactors can generate very complex dynamic phenomena, like, for example, periodic or non-periodic oscillations of temperature and concentration of reagents. In extreme cases, such oscillations may be chaotic.

Considering these properties, I have decided to investigate whether by means of Mandelbrot's method it is possible to obtain complex structures from the solutions of chemical reactor models. The premise enabling the presumption that the obtained pictures may be equally complicated was the fact that chemical reactor models may generate chaos [12]. In particular, chaos is inseparably related to fractals. Thus, such algorithm should be found which could test the impact of the initial values of temperature and concentrations of the reagents on the type of a mathematical solution. In consequence, I obtained the pictures shown in Fig. 6.4. Different shades denote different sensitivity of the model to changes in the initial values. In the original, the fractal is colorful [12,13].

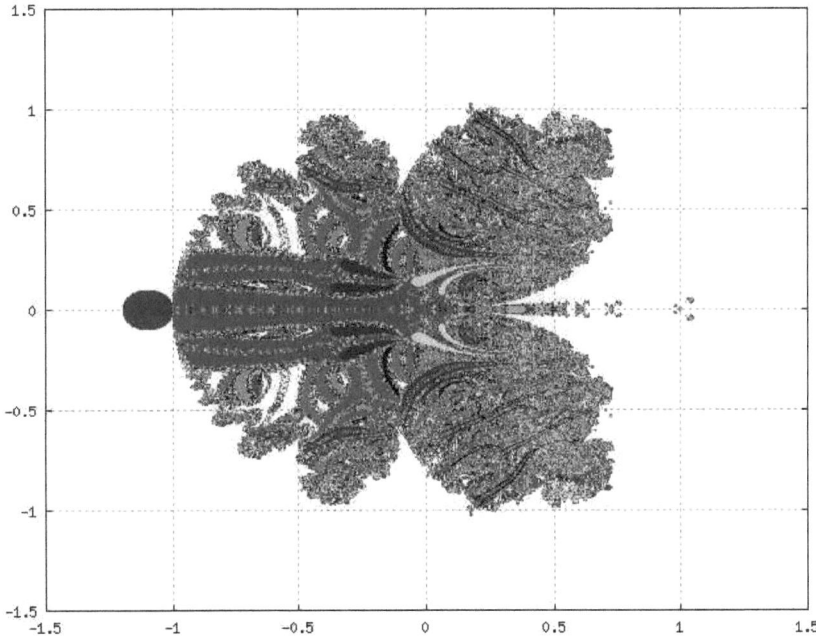

Fig. 6.4. Impact of the initial conditions on the stability of the solutions
of chemical reactor models

To determine the dimension of the structure in Fig. 6.4. I applied the above – mentioned lattice method. To assess if the structure is a fractal, it was necessary to check whether the tested dimension had a value similar to specific fragments of the picture (Fig. 6.5 and 6.6). As a result, it turned out that all the investigated domains had, more or less, the same dimension $D=1.7$, which proved that the structure in Fig. 6.4 is a fractal.

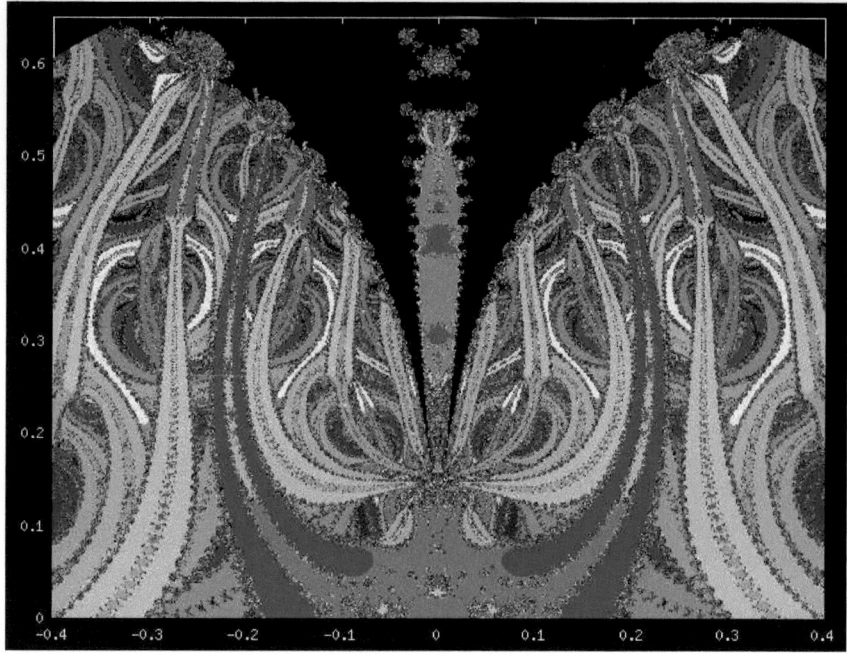

Fig. 6.5. Fragment of Fig. 6.4

Fig. 6.6. Fragment of Fig. 6.5

Now, let us try now to answer the question which is the title of this chapter: What do art and a chemical reactor have in common? Or, otherwise, does a chemical reactor model have something in common with art? It may be stated that the model "paints" pictures.

It follows from Taylor's publication that it is possible, to a bigger or smaller degree, to assess the aesthetic value of a given work of art. This means that aesthetics does not have to be subjective, as mathematics is also not subjective. Anyway, according to Taylor, the most pleasing to human eye is the structure of Pollock's works of the fractal dimension within the range from 1.3 to 1.5. Surely, similar rules also apply to other fields of art, for example, to music.

Just like in the case of the fractal pictures obtained from the calculations of reactor models (Fig. 6.4-6.5), there is also harmony in Pollock's paintings. In his article Taylor remarked: " *Suddenly, the secrets of Jackson Pollock seemed to fall into place for me: he must have adopted nature's rhythms when he painted. At this point, I realized I would have to head back into science to determine whether I could identify tangible traces of those rhythms in his artwork.*" [1].

So, what do art and a chemical reactor have in common? Absolutely nothing. Art is created by man, by the brain, which- as we shall prove below- does not work in terms of algorithms. A real work of art, created by a human being, is unrepeatable. Whereas; the picture generated by a computer can be duplicated any number of times with 100 % accuracy. While creating a work of art, man is often driven by the so called inspiration. It would be difficult to attribute such feature to a machine.

Chapter 7. Is our brain a computer?

The brain, just like a computer, accumulates and transforms information. So, may it be claimed that the brain is just a very complicated computer?

This issue is not as blank as it seems, and, in consequence, it is derived from mathematics and the theory of algorithms to philosophical questions.

David Hilbert, German mathematician, posed the following thesis: if mathematics is a set of strictly defined rules, is it at all possible to create a universal automatic machine (automata, algorithm, program) based on the rules to solve any mathematical problems, for instance, prove theorems. Austrian mathematician, Kurt Gödel, presented a certain theorem, which – in its general sense- sounds as follows: for any self- consistent system of rules there are theorems that cannot be proved by means of this rules. This theorem, no wonder, was a blow to Hilbert's thesis, because, if it is impossible to prove all mathematical theorems, there is no general automata that could prove it.

Let us look, however, closer at this unusual issue. Roger Penrose conducted an following interesting reasoning [2,3]. To do this, let us formulate a theorem- for practical purposes, let us call it "G". G states: "there is no proof of G". This means, that G cannot prove what it states! Thus, what remains is to resolve is if the sentence: "G states that G cannot be proved" is true or false. Otherwise, if G is true or false. Let us assume, just for the moment that G is false, and , accordingly- against what it wants to demonstrate us- there is a proof of G. This would mean that G states what is false and there is a proof of the non-existence of such proof! This is an explicit contradiction. Thus, it is possible to refute the assumption that G is false. Accordingly, there is no choice but to accept that G is true. This means that we are certain of the truth of something that we cannot prove! Although the question how we know it, even if we cannot prove it, remains open.

So, what is the connection between theorem G and the thesis posed by Hilbert? Certainly, theorem G refutes Hilbert's thesis, as it makes us aware that there are appropriate mathematical rules that cannot be proved by any mathematical rules (within the same system). In consequence, it is not possible to create a general automatic machine based on these rules to solve any mathematical problem. Hence, what does this have in common with the computer and the human brain? Well, the computer is a machine that accomplishes only and solely strictly defined algorithms. Thus, the computer (or, should it rather be said the algorithm completed by it) will never be capable of proving the validity of theorem G!, as it does not have anything else than a set of mathematical rules at its disposal- and, as mentioned before, they are not sufficient to prove if G is true. Hence, the computer will never find out if G is true. On the other hand, we, as humans, know this because we understand it. Because the knowledge of G being true cannot be reached by means of algorithms, the conclusion is that the human brain does not work and comprehend the surrounding world in an algorithmic manner! The brain is not a computer, even immeasurable. The computer does not understand anything, but the brain does. The computer does not have a consciousness, but the brain does. It is difficult to not agree with this conclusion [2,3].

At this point, we approach, in a certain sense, the issue of artificial intelligence. There is only one intelligence, involving the consciousness; however, its performance may be artificial or true (which has no relevance to memory and the ability to remember). The true intelligence should be understood as the intelligence contained in living organisms. The artificial intelligence should be understood as the intelligence contained in a machine, thus, in the algorithm that it completes.

As stated before, the computer is not capable of comprehending what a living organism knows without any need to use algorithms. Thus, it is impossible to accomplish intelligence in a machine! As stated above, each algorithm is limited, which is surely sustained by its absence of the consciousness of the validity of theorem G. On the other hand, the brain has this consciousness, so it has a higher position in the hierarchy of the possibility of cognition. But, is it limitless? According to Gödel's theorem, it seems not the case. It knows that G is true, but in the surrounding cognition opportunities it is not capable of crossing another threshold of consciousness. Therefore, in accordance with G, it will never comprehend itself! Likewise, it will never be possible to prove the mathematical theorem of G, only by the means of a set of mathematical rules, and the brain will never be capable to understand its actions. In either case, there is no external support point, as in the famous sentence of Archimedes: "give me a place to stand and I will move the world". In proving theorem G this point is the superior set of rules reaching beyond the set of mathematical rules. The superior consciousness is essential to understand the activity of the human brain. A similar conclusion was reached by E. Alexander [14].

To have a more profound insight into the fact that the brain is incapable of comprehending its operation, let us recall a specific problem raised by Alan Turing, English mathematician, who formulated the theorem which states that there is no universal algorithm which can determine that any other algorithm could generate the final results, i.e. complete its work (*the halting problem of Alan Turing*). According to this theorem, there is no computer that can comprehend and control the operation of any other computer.

It is also worth-mentioning at this point that the brain operates in a very intriguing manner. Unlike a computer, while registering the surrounding real world, the brain is not focused on details, but on transforming pictures as a whole. A good example is the ease of reading the text consisting of words, in which letters are written at random. For instance: " biran". The condition is that the first and the last letters are at their proper place. The next paragraph was written in this way. The letters in the words were scrambled by means of the logistic chaotic equation:

$$x_{k+1} = rx_k(1 - x_k).$$ (7.1)

To obtain this representation, the following procedure was followed:
- The number of recurrence steps corresponded to the number of letters in the word, with the exclusion of the first and the last letter;
- The values of variable x were sorted out in accordance with an increasing relation, which changed the order of letters in the word.

Thus, it may be stated that human brain can read the text which, in fact, does not exist!

The albiity of such rnadeig is, ubdnotldeuy, a piiovste fuaetre of our brian. It eabelns btteer unnrndtdsieag of the txet. Hwveeor, it can also be dragoneus. It may occur that our biran slhal fcore us to irtreenpt seitonhmg, wchih, in fact, neevr hppneead. This may be prlltcaiuray dragonues under ctaeirn cseitnurcacms, for empxlae: in crout tieoisetmns, wehn the wetsins is cncivnoed abuot sneeig the eenvt taht has ralley nveer tkaen pclae.

It should be added that the case in which the internal letters in the word are not mixed, but ordered in an inverted manner, is very difficult to decipher, for instance: "ceupmtor". Yet, the most difficult to decipher are the words in which the internal letters are in the right order, but the first and the last letters are, respectively, in their wrong position, for instance: " lrtificiaa".

Chapter 8. Problems with alien civilization

Among scientists there are some who support the so called: strong artificial intelligence. They believe that in a short time computers shall outrival humans as far as intellectual abilities are concerned. This is certainly in compliance with the assumption that human brain is just a computer. Such concept is called *computer functionalism.* According to this idea human mind is a program installed in the brain.

The assumption that rapid advancement of computers (algorithms) shall exceed the intellectual level of the human brain leads, in consequence, to the acknowledgment that there may be non-biological life and civilization in the Universe. Such civilization could exist on planets hostile to biological life, for example on some object devoid of any atmosphere, water, etc. The subsistence of such civilization would only depend on the availability of resources required to construct machines and energy necessary for their operation (for instance: radiation energy from the neighboring star). Such things are abundant in the Universe. So, while searching for extraterrestrial civilization, why do people ingenuously check if there are appropriate life-sustaining biological conditions on a given planet? Machines do not need such conditions.

Concurrently, it should be noticed that such *"creatures"* would never die. Only their mechanisms would use up, but their consciousness could be transferred, at any time, to other newer generation species. Consequently, cloning would be a natural procedure. On principle, one representative of such creatures endowed with the consciousness would suffice, the rest being only mindless slave machines controlled by the representative to sustain its own existence.

Does this sound absurd? Science fiction? No. It is just a consequence of the concept of the advancement of artificial intelligence.

Another problem is the conviction that human beings would be able to communicate with representatives of alien civilization, no matter if it was biological or computer one. It is merely sufficient to admit that people cannot communicate with animals. This, in turn, could be very dangerous to us- people. Hence, the hope that alien extraterrestrial civilization would be able to show us how to solve problems that have so far been insoluble, may be utterly naïve.

Likewise, it cannot be excluded that alien civilization would possess superior consciousness that would make it possible to understand the human brain. Then, the relation between the two species would be: people versus algorithms. Man is capable of understanding how an algorithm works, but the algorithm is not able to comprehend how human brain works. An algorithm is not even aware of its own existence. Although we, as people, are aware of our existence, which makes us different from algorithms, we may still be unaware of the existence of a higher form of intelligence that surrounds us.

Chapter 9. Theory of relativity for everyone

As we have touched upon the issue of extraterrestrial civilization, it is worth considering what happens in the Universe. What occurs in the Universe of high speeds and huge distances? Well, what takes place there is completely different from what we are used to here, on Earth. Speeds do not add up, the passage of time is not the same, the length of the same objects is not identical. Absurd? No! Completely natural physical phenomena discovered by Einstein and described in his detailed and general theory of relativity. Although very simple, this theory is hardly understandable. It is incomprehensible, because it is strange to our senses. How to understand that some object is one meter long, here, on Earth, but somewhere else it is only half a meter long or has even the length of zero? Such circumstances are similar to the situation where we try to imagine, for example, the fourth dimension, but it is not accessible to our senses. However, this is not the reason to claim that the fourth dimension does not exist or that it is something absurd. The string theory requires the existence of even ten dimensions. Likewise, the theory of relativity.

The bases of the theory of relativity are the assumptions that the speed of light is limited and cannot be exceeded, and that light always travels at the same speed c. The remaining parts of the theory are just the consequences of these assumptions.

At first, let us consider why is the speed of light limited, and not infinitely high? To answer this question we shall examine the following example.

At the cross roads, before the red light, there is a line of trucks. Let us assume that only the driver of the first truck can see the traffic lights. At a certain moment, the lights change from red to white. What happens? As the change of the lights is visible only to the driver of the first car, he starts driving. When will the second car start to move? Only after its driver notices that the first car has moved on. The third truck will start when its driver can see that the second car has started to move on, etc. So, the conclusion is that all the trucks will not start moving at the same time, because they must receive the information that driving can be started, which – as shown in this example requires time. This means that the flow of information cannot run at indefinitely high speed.

In the physical reality that surrounds us, it is light that moves at the highest speed. This means that no information can be sent faster than with the speed of light.

The conclusion: the speed of forwarding information, and, at the same time, the speed of light, must be limited. The fact that it equals to 300000 km/s is of secondary importance.

In physics, the smallest distance that has any physical meaning is the so called Planck's length $l_p = 1.616199(97) \times 10^{-35}$ [m]. Whereas, the smallest unit of time that has any physical meaning is Planck's time $t_p = 5.39106(32) \times 10^{-44}$ [s]. This is the time required for a photon to move to the distance equal to Planck's length. Hence, it is possible to calculate the speed of the movement of the light photon: $c = l_p / t_p \approx 300000$ [km/s].

As far as the constancy of the speed of light is concerned, let us perform a certain intellectual speculation. So, let us imagine that in the nth dimensional space (n>3) the values of this speed change in a continuously unlimited way (Fig. 9.1(A)). However, what we can see in our Universe is just a projection (view) on our three-dimensional space (Fig. 9.1(B)). This projection is such that the curving of the line in space *n-D* finishes at point *c=300000km/s* in 3-D space. But, this is just a speculation.

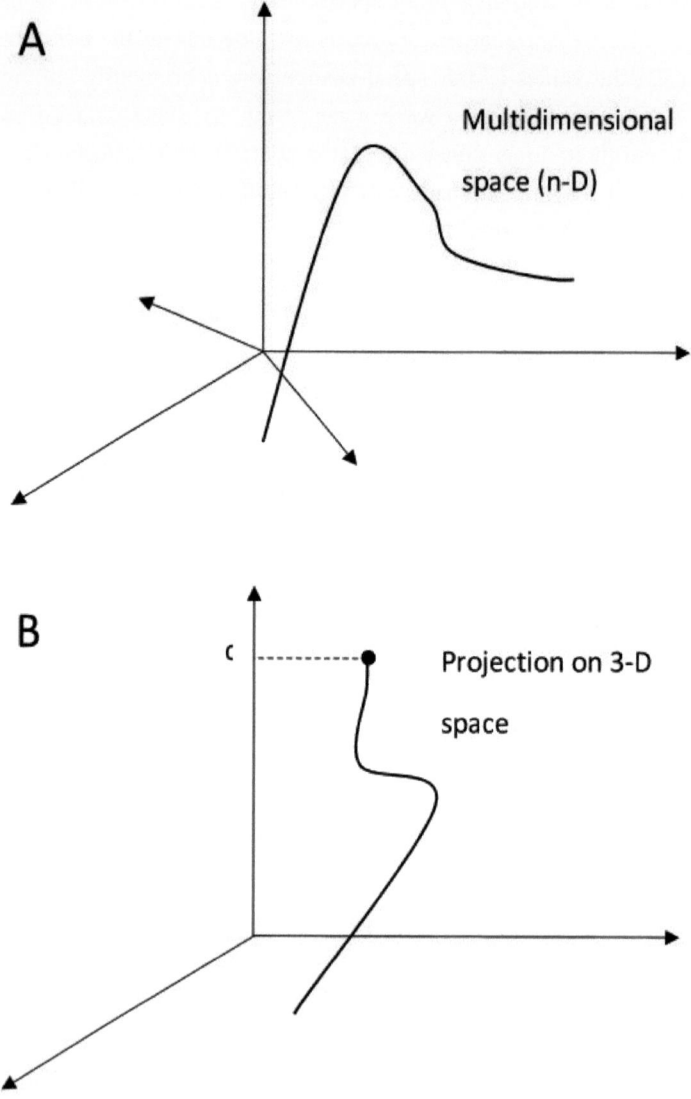

Fig.9.1. Projection on space

Let us now think of the consequences of the steadiness of the speed of light by examining a certain intellectual experiment. There is a rocket in space travelling with the speed of light c. Inside the rocket there is a person that turned on the torch. What shall we see on the Earth? We can see that the light does not distance from the torch. If it did, it would have to travel, in relation to us, with the speed higher than c, which is impossible (Fig. 9.2).

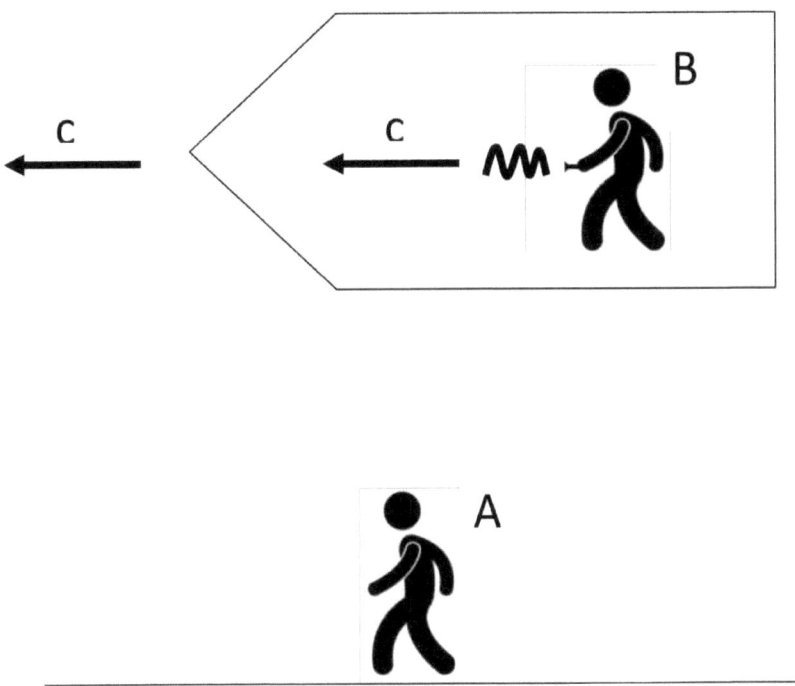

Fig. 9.2. According to observer A the time in the rocket does not move forward

The fact that the light does not move away from the torch in the rocket travelling, in relation to us, at speed c, denotes that inside it no changes take place. As a matter of fact, time is nothing else than a measure of change, so, according to our observations, the time inside the rocket has stopped.

If the rocket slowed down a little bit, then – according to our observations- the light would slowly move away from the torch, and, in consequence, changes would take place in the rocket, i.e. the time would flow.

If the light moved away from the rocket to the distance d_R, then it would also move away from us to the distance much bigger, equal to $d_Z > d_R$. Accordingly, we would be able to observe more changes on the Earth than in the rocket, which, in turn, would signify that although the time in the rocked did not stop, but flew slower than on the Earth. In other words, the passengers of the rocket, and the entire system within, would get older slower than here on the Earth.

At this point, someone could pose the question: What does it mean, in practice, that the time in the fast moving rocket flows slower than on the Earth or that is has even stopped? How does the pilot of the rocket feel about this? The answer is explicit: he feels nothing. The time in the rocket flows slower or has stopped only in accordance with our observations. The pilot of the rocket does not notice this. Conversely, he thinks that the time has stopped on the Earth, because it is the Earth that moves away from the rocket with the speed of light.

So, who is right: we on the Earth or the pilot of the rocked travelling in relation to us? For both parties, this is a description of reality only from the point of view of their own system. The symmetry breaks down after a physical meeting of both parties.

Let us assume that the pilot of the rocket has made a decision to enter the Earthen system and meet his twin brother. He did this exactly when ten years had passed on the Earth. It is obvious that after the meeting their statements must be identical and the interpretation of the Earthen twin must oblige. The fact that his brother has left the rocket and has come back to the Earth shall not evoke any changes on the Earth. In other words, the pilot of the rocket has to acknowledge the binding rules of the system in which he is now. Accordingly, his Earthen brother has to be older than the other twin.

If the Earthen twin decided to leave the Earth and enter the dashing rocket, he would then be younger than his twin brother (Fig. 9.3).

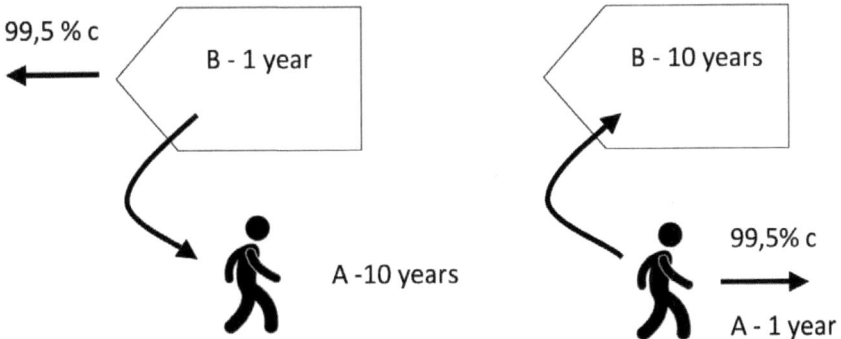

Fig. 9.3 *"The twin paradox"*

To recapitulate, the principle of the so called *"twin paradox"* is quite simple: when we move to a given system of reference, we are obliged by the rules that hold in this system. Likewise, while entering a foreign country we have to abide by the rules of this country, and not otherwise.

Extreme cases are intriguing. Let us assume that the rocket is dashing with the speed of light in relation to the Earth. The pilot left the rocket and met with his twin brother at the moment when the Earthen clock moved on, for example, by one year. As we already know, it will turn out that the twin from the rocket will be younger. But what is really interesting is the fact that the rocket twin shall not grow older. He will also not be aware that he has travelled at all, not to mention that his journey has been made with the speed of light. Undoubtedly, he will be surprised that his Earthen twin has grown older in the twinkling of an eye.

Another extreme case is the situation when after certain passage of time on the Earth, for example: one year, the Earthen twin decides to enter the rocket. What will he claim? We should remember that he has to acknowledge the rules that oblige in the rocket. Let us also keep it in mind that no matter how much time will pass in the rocket, no changes will take place on the Earth (time does not flow forward). In relation to the rocket, the Earth moves with the speed of light. In parallel, it is beyond any doubt that the Earthen twin is one year older, as he made the decision to enter the rocket after such period of time. The answer is that he will not meet his brother anymore (Fig. 9.4).

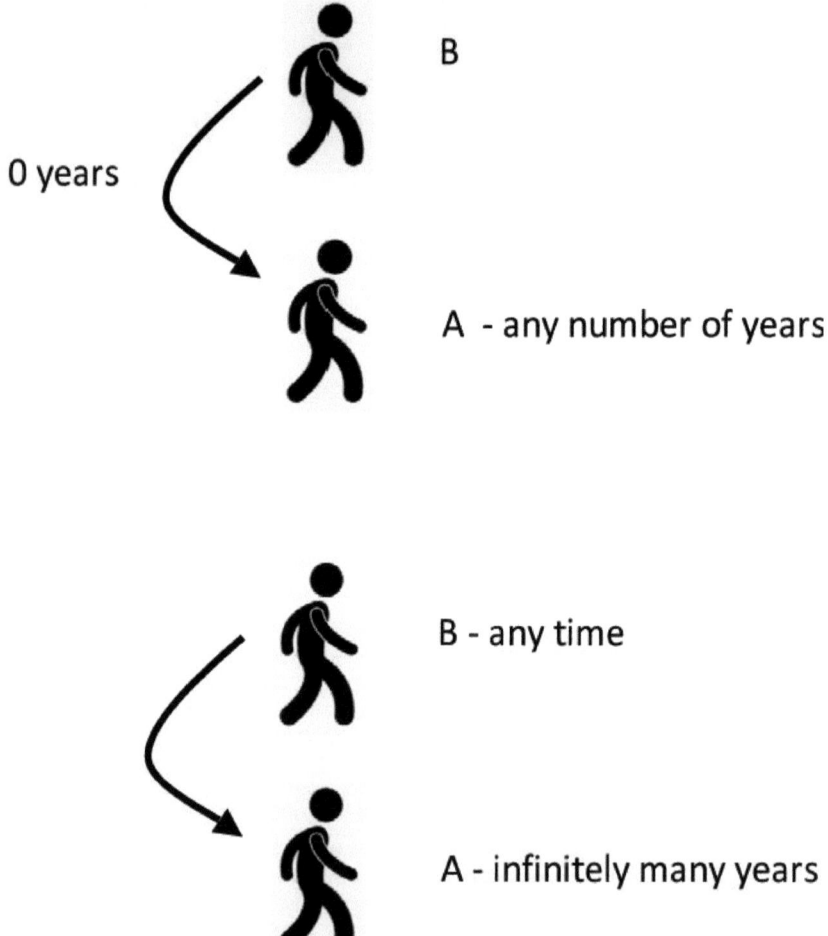

Fig. 9.4. *"The twin paradox"* under extreme conditions

Anyway, the phenomenon described above, labeled in professional literature as: "*the twin paradox*" is by no means a paradox.

Let us consider at this point the concept of time, which is nothing else but a measure of changes, just like temperature is a measure of heat. Yet, they are not physical quantities (values).

Let us assume that there is a source of light emitting rays in all directions and located in a train which moves along the platform. An observer standing on the platform carries out the following experiment. He calculates N quanta of the distance made by the light ray receding from the platform while the train moved to k quanta distance. For the observer the light ray moved from the source by the distance equal to $(N-k)$ quanta in the direction coincident with the movement of the train and $(N+k)$ quanta in the opposite direction. In each case, this means that when the train moves along the platform, the number of the changes in the train is different from the number of the changes on the platform. Accordingly, a question arises whether the number of the changes is bigger, in the train or on the platform. First and foremost we should realize that the number of the changes must be equal in all parts of the train. The proportionality of N towards $(N-k)$ as well as towards $(N+k)$ ($N = \alpha(N-k); (N+k) = \beta N; \alpha > 1; \beta > 1$) leads to the following relation: $\alpha \dfrac{N-k}{N} = \beta \dfrac{N}{N+k}$. This means that $N = \sqrt{\dfrac{\alpha}{\beta}}\sqrt{(N^2 - k^2)}$. Hence, it may be concluded that from the observer's point of view - N changes occurring on the platform correspond to $\sqrt{(N^2 - k^2)}$ changes occurring in the train moving along the platform. Because $N > \sqrt{(N^2 - k^2)}$, the observer situated outside the train concludes that the number of the changes in the platform is bigger than the number of the changes in the moving train.

As the two systems (the train and the platform) are mutually symmetrical (none is privileged), the observer located inside the train will conclude that the platform is in motion and the number of the changes in the train is bigger than the number of the changes on the platform.
Let us assume that there is a ruler in the train the distance of which we want to measure. The measuring manner is assumed to be a number of N quanta to be

traveled by the ray along the ruler. It follows that its length at rest (when the train is standing on the platform) is $l_o = N$ quanta. Furthermore, let us assume that the train and the ruler move along the platform and that the ruler is located parallel to the moving direction. As already mentioned, from the external observer's point of view, a certain number of the changes on the platform corresponds to a smaller number of the changes in the moving train. Accordingly, from the observer's point of view- if the ruler on the platform was measured in N quanta steps, in the moving train it will be measured in $\sqrt{N^2 - k^2}$ quanta. So, the external observer concludes that the length of the ruler measured in the train is $l = \sqrt{N^2 - k^2}$ quanta, which, in combination with the length at rest leads to relation $l = l_o \dfrac{\sqrt{N^2 - k^2}}{N}$. For the observer located on the platform the length of the ruler in the moving train is smaller than the length of the same ruler at rest in relation to the platform.

At this point we may introduce the concept of time as a measure of changes. Thus, let us define Δt_o as the passage of time on the platform, whereas Δt will denote the passage of time in the moving train. Accordingly, we derive relations: $\Delta t_o = \dfrac{1}{c} N$, $\Delta t = \dfrac{1}{c} \sqrt{N^2 - k^2}$, where $\dfrac{1}{c}$ is the proportionality coefficient related to the movement of light (c shall denote the velocity of light). From such simple relations the following equation follows: $\dfrac{\Delta t}{\Delta t_o} = \dfrac{\sqrt{N^2 - k^2}}{N}$. So, the external observer may conclude that there is a different measure in the moving train and that the passage of time in the train is shorter than the passage of time on the platform.

In our experiment the number of N changes of the light moving along the platform corresponds to k changes of the moving train. Accordingly, if for the observer located on the platform the measure of the light ray displacement of N quanta is the time interval $\Delta t_o = \dfrac{1}{c} N$, the measure of the displacement of the train of k quanta is the same range $\Delta t_o = \dfrac{1}{v} k$, where $\dfrac{1}{v}$ is the proportionality coefficient related to the movement of the train along the platform (v shall denote the velocity of the train moving along the platform). It follows from the two relations that $\dfrac{N}{k}=\dfrac{c}{v}$. By substituting this relation in the previously discussed equations we arrive at $\Delta t = \Delta t_o \sqrt{1-\left(\dfrac{v}{c}\right)^2}$ and $l = l_o \sqrt{1-\left(\dfrac{v}{c}\right)^2}$.

To issues still remain two clarify. Firstly, why in the discussed experiment was it not merely assumed that $k=1$? Such approach would be erroneous, as the relation $\dfrac{k}{N}$ may be any positive rational number smaller than 1. Let us assume as an example that this relation is $\dfrac{5}{6}$. Then, the mathematical sequence in the successive steps is the following: $\left(\dfrac{1}{1.2}\right)$, $\left(\dfrac{2}{2.4}\right)$, $\left(\dfrac{3}{3.6}\right)$, $\left(\dfrac{4}{4.8}\right)$, $\left(\dfrac{5}{6.0}\right)$ etc. As the quanta steps may be only integral numbers, the real sequence is:
$\left(\dfrac{k}{N}\right)=\left(\dfrac{1}{1}\right),\left(\dfrac{2}{2}\right),\left(\dfrac{3}{3}\right),\left(\dfrac{4}{4}\right),\left(\dfrac{5}{6}\right)$ etc. The external observer will therefore conclude that there are no changes in the moving train in the first step of the experiment, and, using the concept of time, the observer will state that time does not pass in the train.

The second issue concerns the measurement of objects length when they are exactly equal to Planck lenght. Let us assume that the object is moving at nearly the speed of light, and the task of an external observer is measuring its length. According to the Special Theory of Relativity the length should be shortened to size smaller than the Planck length. Therefore there is a conflict between the concept of the Planck length, and the Special Theory of Relativity. The method presented in this paper solves this problem.

The physical object having a non-zero mass at rest, maybe move relative to an external observer, at most of about $k=N-1$ quantum steps ($k = N$ means that the object moves like light). This means that - from the point of view of an external observer - the length of such an object is measured in increments of $\sqrt{N^2-k^2}=\sqrt{2N-1}$ quantum steps. Relativistic length equal to $l=l_o\dfrac{\sqrt{2N-1}}{N}$. If, therefore, an object at rest, has a length equal to the Planck length (equal to one quantum step - $N = 1$), so when moving it has the Planck length as well.

Relativistic length equal to length of rest and is $l = l_o \dfrac{\sqrt{2N-1}}{N} = l_o$.

It should also be indicated that the phenomenon of the passage of time and dissimilar aging does not only concern moving objects, since time flows in the vicinity of mass in a manner different from the circumstances of empty space. This is all in compliance with the general theory of relativity.

It follows from the theory of relativity that the light is gravitationally drawn by every object endowed with mass. In its vicinity, the trajectory of the light begins to curve and, in consequence, the ray shot from point A in the direction of point B reaches point C (curve "a" in Fig. 9.1).

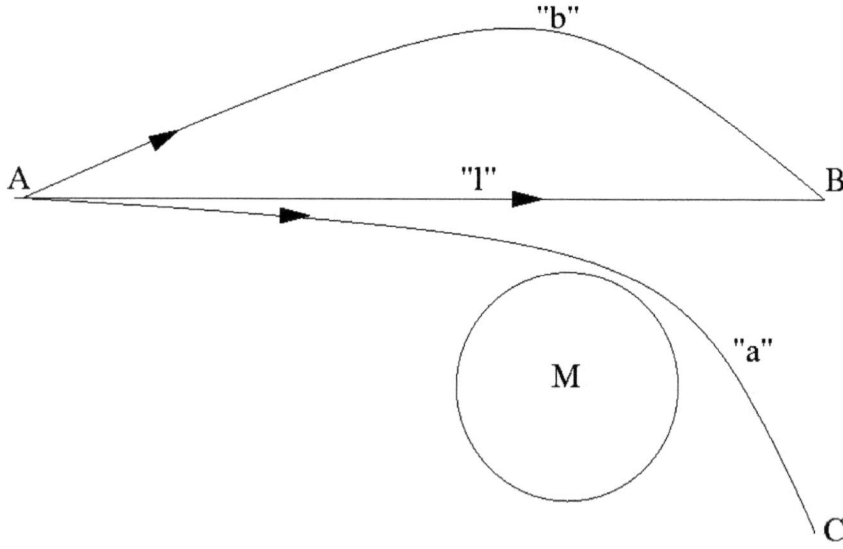

Fig. 9.1. Curving of the light trajectory

So, what should be done to make the ray sent from point A reach point B? Obviously, it should be shot at such angle that, even after being curved by the mass, it would land at B ("curve b").

If there was no mass in the vicinity of the ray trajectory, the ray would not be curved, and the time of relocation from A to B along straight line "l" would be equal to:

$$t_l = l/c \qquad\qquad (9.1)$$

c - speed of the light.

As curved trajectory b is longer than section l $(b>l)$, time t_b is longer than time t_l

$$t_b = \frac{b}{c} > t_l = \frac{l}{c}. \qquad\qquad (9.2)$$

The shooting of the ray from point A and its reaching point B, are two different events. The difference between them proves the change that has occurred in the system. The change takes more time if mass is present in the vicinity of the trajectory, because then the journey from A to B takes longer time $(t_b > t_l)$. And because time is a measure of changes, the conclusion is that time flows slower in the vicinity of mass.

The bigger the mass is, the stronger its drawing of the light ray, which results in a bigger curve of the trajectory of its journey. Accordingly, the bigger the mass M, the bigger the angle at which the ray should be shot from point A to reach B. This, in turn, leads to the elongation of trajectory b, and, consequently, to an increase of journey time t_b. Hence the conclusion that the bigger the mass, the slower the passage of time in its vicinity. In an extreme case, if the mass is so huge that the ray has to be shot at angle 90°, it is obvious that it will never reach point B. This means that in the vicinity of such huge mass, time completely stops. Such types of mass are called the black holes. There is no way for the light ray within their range to be released and reach anywhere outside the black holes.

Epilogue. The power of number *1*

Let us reflect on the value of number *1* raised to any power *x*? It is always 1? Well, not always. Number *1* raised to certain power may have very many different values, sometimes even infinitely many.

For example, number *1* raised to the power of 1.5 has two different values, raised to the power of 0.25 has four different values, raised to the power of 0.415 has as many as 200 different values , whereas, raised to the power of $\sqrt{2}$ has an infinite number of values.
How is that possible? Let us analyse what number 1 really is.

It fulfils the equation:

$$1 = e^{2\pi k i}$$ (10.1)

where k is any integral number. The right side of the equation has a complex form, which may be represented by means of cosine and sine in the following way:

$$e^{2\pi k i} = \cos(2\pi k) + i\sin(2\pi k).$$ (10.2)

Making use of the above equations, we can find out what is the value of 1 raised to any power *x*:

$$1^x = \left(e^{2\pi k i}\right)^x = e^{2\pi k x i} = \cos(2\pi k x) + i\sin(2\pi k x).$$ (10.3)

Let us check the value of number 1 raised, for example, to the power 1.5=3/2. Well,

$$1^{3/2} = \cos(3\pi k) + i\sin(3\pi k) \qquad (10.4)$$

has two different values: +1 and –1.
What then is the value of 1 raised to the power 0.25=1/4?

Well,

$$1^{1/4} = \cos(\pi k/2) + i\sin(\pi k/2) \qquad (10.5)$$

has ma 4 different values: +1, *i, -1, -i*.

Likewise, it is easy to check that number *1* raised, for example, to the power 0.415=83/200, has 200 different values.

Generally, if x is a rational number, i.e. it can be represented by two integral numbers m and n in the form $x = m/n$, then, number *1* raised to the power of x has:

- One real value equal to +1 and *(n-1)* different complex values, if n is an odd number,
- Two real values equal to +1 and –1 and *(n-2)* different complex values, if n is an even number.

But, if x is an irrational number, i.e. it cannot be represented by means of two integral numbers (for example: $\sqrt{2}$), number *1* raised to the power of x has one real value equal to +1 and infinitely many different complex values.

References

1. R. P. Taylor. *Order in Pollock's Chaos,* Scientific American, December 2002.
2. R. Penrose. *The Emperor's New Mind,* Oxford University Press, SBN 0-19-851973-7, 1989.
3. R. Penrose. *Shadows of the Mind,* Oxford University Press, ISBN 0-19-853978-9, 1994.
4. M. Berezowski, *Analysis of dynamic solutions of complex delay differential equation exemplified by the extendet Mandelbrot's equation.* Chaos, Solitons & Fractals; **24**/5, pp. 1399-1404, 2005.
5. K. Morawetz. *Bifurcation in kinetic equation for interacting Fermi systems.* Chaos;13(2):572, 2003.
6. H.-O. Peitgen, H. Jürgens, D. Saupe. *Fractals for the Classroom.* Springer-Verlag New York, Inc. 1992.
7. M. Berezowski. *Phase trajectories of a certain mechanical system.* Far East Journal of Dynamical Systems; **13**/1, 85-96, 2010.
8. M.V. Berry & Z.V. Lewis. *On the Weierstrass-Mandelbrot Function.* Proc. Roy. Soc. London Ser.; A 370: 459-484, 1980.
9. P. Du Bois-Reymond. *Versuch einer Classification der willk urlichen Functionen reeller Argumente nach ihren Aenderungen in den kleinsten Intervallen.* J. Reine Angew. Math.; 79: 21–37, 1875.
10. A. Einstein. *Über die von der molekularkinetischen Theorie der Wärme geforderte Bewegung von in ruhenden Flüssigkeiten suspendierten Teilchen.* Annalen der Physik; 17: 549–560, 1905.
11. M. Smoluchowski. *Zur kinetischen Theorie der Brownschen Molekularbewegung und der Suspensionen.* Annalen der Physik; 21: 756–780, 1906.
12. M. Berezowski, *Fractal solutions of recirculation tubular chemical reactors*, Chaos, Solitons&Fractals, **16**, 1-12, 2003.
13. M. Berezowski, *Fractals galery*, http://c504c.skroc.pl
14. E. Alexander. *Proof of Heaven, The New York Times Best Seller, 2012.*